Contents

621.3815 TOO

28783.

Preface	vi	
A word about safety	vii	
1 Electrical fundamentals	1	
2 Passive components	18	
3 D.C. circuits	48	
4 Alternating voltage and current	68	
5 Semiconductors	83	
6 Power supplies	109	
7 Amplifiers	121	
8 Operational amplifiers	146	
9 Oscillators	160	
10 Logic circuits	172	
Appendix 1 Student assignments	190	
Appendix 2 Revision problems	193	
Appendix 3 Answers to problems	202	
Appendix 4 Semiconductor pin connections	204	
Appendix 5 Decibels	206	
Index	208	

Electronic Circuits Student Handbook

Electronic Circuits Student Handbook

Michael Tooley, BA
Dean of the Faculty of Technology
Brooklands College

Newnes
An imprint of Butterworth-Heinemann Ltd
Linacre House, Jordan Hill, Oxford OX2 8DP

$\mathcal{R}$ A member of the Reed Elsevier plc group

OXFORD   LONDON   BOSTON
MUNICH   NEW DELHI   SINGAPORE   SYDNEY
TOKYO   TORONTO   WELLINGTON

First published 1995
© Michael Tooley 1995

British Library Cataloguing in Publication Data
A catalogue record for this book
is available from the British Library.

ISBN 0 7506 2118 4

Library of Congress Cataloging in Publication Data
A catalogue record for this book is
available from the Library of Congress.

Typeset by Graphicraft Typesetters Ltd, Hong Kong
Printed and bound in Great Britain by
Martins the Printers Ltd, Berwick upon Tweed

Preface

This book has been designed to help you understand how electronic circuits work. It provides the basic under-pinning knowledge necessary to appreciate the operation of a wide range of basic electronic circuits including amplifiers, logic circuits, power supplies and oscillators.

The book is ideal for students following formal courses (e.g. GCSE, GNVQ, BTEC, City and Guilds, RSA, etc.) in schools, sixth-form colleges, and further/higher education colleges. It is equally well suited to those who may be returning to study or who may be studying independently.

Worked examples have been liberally included within the text in order to give you an appreciation of the solution of simple numerical problems related to the operation of basic circuits. In addition, a number of problems can be found at the end of each chapter. These can be used to check your understanding of the text and to give you some experience of the 'short answer' questions used in BTEC, RSA, City and Guilds, and other in-course assessments. For good measure, we have also included 50 revision problems in Appendix 2.

To satisfy a representative selection of the 'evidence indicators' required by the GNVQ awarding bodies (currently BTEC, City and Guilds and RSA), we have also included twelve sample coursework assignments. These should give you plenty of 'food for thought' as well as offering you some scope for further experimentation.

While the book assumes little previous knowledge you need to be able to manipulate basic formulae and understand some simple trigonometry to follow the numerical examples. A study of mathematics to GCSE level (or its equivalent) will be more than adequate to satisfy this requirement. A scientific calculator will also be useful when tackling the problems. Each chapter ends with a summary of the formulae and circuit symbols introduced within the chapter.

In the later chapters of the book, a number of representative circuits (with component values) have been included together with sufficient information to allow you to adapt and modify the circuits for your own use. These circuits can be used to form the basis of your own practical investigations or they can be combined together in more complex circuits.

Finally, please don't hesitate to put your new fund of knowledge into practice. A great deal can be learned by building, testing and modifying simple circuits. To do this you will need access to a few basic tools and some minimal test equipment. Your first purchase should be a simple multi-range meter, either digital or analogue. This instrument will allow you to measure the voltages and currents present in your circuits and compare them with predicted values. If you are attending a formal course of instruction and have access to an electronics laboratory do make full use of it!

A word about safety

When working on electronic circuits, personal safety (both yours and of those around you) should be paramount in everything that you do. Hazards can exist within many circuits – even those that, on the face of it, may appear to be totally safe. Inadvertent misconnection of a supply, incorrect earthing, reverse connection of a high-value electrolytic capacitor, and incorrect component substitution can all result in serious hazards to personal safety as a consequence of fire, explosion or the generation of toxic fumes.

Potential hazards can be easily recognized and it is well worth making yourself familiar with them so that pitfalls can be avoided. The most important point to make is that electricity acts very quickly; you should always think carefully before taking any action where mains or high voltages (i.e. those over 50 V, or so) are concerned. Failure to observe this simple precaution may result in the very real risk of electric shock.

Voltages in many items of electronic equipment, including all items which derive their power from the a.c. mains supply, are at a level which can cause sufficient current flow in the body to disrupt normal operation of the heart. The threshold will be even lower for anyone with a defective heart. Bodily contact with mains or high-voltage circuits can thus be lethal. The most severe path for electric current within the body (i.e. the one that is most likely to stop the heart) is that which exists from one hand to the other. The hand-to-foot path is also dangerous but somewhat less dangerous than the hand-to-hand path.

Before you start to work on an item of electronic equipment, it is essential not only to switch off but to disconnect the equipment at the mains by removing the mains plug. If you have to make measurements or carry out adjustments on a piece of working (or 'live') equipment, a useful precaution is that of using one hand only to perform the adjustment or to make the measurement. Your 'spare' hand should be placed safely away from contact with anything metal (including the chassis of the equipment which may, or may not, be earthed).

The severity of electric shock depends upon several factors including the magnitude of the current, whether it is alternating or direct current, and its precise path through the body. The magnitude of the current depends upon the voltage which is applied and the resistance of the body. The electrical energy developed in the body will depend upon the time for which the current flows. The duration of contact is also crucial in determining the eventual physiological effects of the shock. As a rough guide, and assuming that the voltage applied is from the 250 V 50 Hz a.c. mains supply, the following effects are typical:

Current	Physiological effect
less than 1 mA	Not usually noticeable
1 mA to 2 mA	Threshold of perception (a slight tingle may be felt)
2 mA to 4 mA	Mild shock (effects of current flow are felt)
4 mA to 10 mA	Serious shock (shock is felt as pain)
10 mA to 20 mA	Motor nerve paralysis may occur (unable to let go)
20 mA to 50 mA	Respiratory control inhibited (breathing may stop)
more than 50 mA	Ventricular fibrillation of heart muscle (heart failure)

It is important to note that the figures are quoted as a guide – there have been cases of lethal shocks resulting from contact with much lower voltages and at relatively small values of current. It is also worth noting that electric shock is often accompanied by burns to the skin at the point of contact. These burns may be extensive and deep even when there may be little visible external damage to the skin.

The upshot of all this is simply that **any potential in excess of 50 V should be considered dangerous**. Lesser potentials may, under unusual circumstances, also be dangerous. As such, it is wise to **get into the habit of treating all electrical and electronic circuits with great care.**

1

Electrical fundamentals

This chapter has been designed to provide you with the background knowledge required to help you understand the concepts introduced in the later chapters. If you have studied electrical science or electrical principles beyond GCSE or GNVQ Intermediate Level then you will already be familiar with many of these concepts. If, on the other hand, you are returning to study or are a newcomer to electrical technology this chapter will help you get up to speed.

Fundamental units

The units that we now use to describe such things as length, mass and time are standardized within the International System of Units. This SI system is based upon the seven **fundamental units** (see Table 1.1).

Derived units

All other units are derived from these seven fundamental units. These **derived units** generally have their own names and those commonly encountered in electrical circuits are summarized in Table 1.2 together with the physical quantities to which they relate.

If you find the exponent notation shown in the table a little confusing, just remember that V^{-1} is simply $1/V$, s^{-1} is $1/s$, m^{-2} is $1/m^2$, and so on.

Example 1.1

The unit of flux density (the tesla) is defined as the magnetic flux per unit area. Express this in terms of the fundamental units.

Solution

The SI unit of flux is the weber (Wb). Area is directly proportional to length squared and, expressed in terms of the fundamental SI units, this is square metres (m^2). Dividing the flux (Wb) by the area (m^2) gives Wb/m^2 or $Wb\ m^{-2}$. Hence, in terms of the fundamental SI units, the tesla is expressed in $Wb\ m^{-2}$.

Example 1.2

The unit of electrical potential, the volt (V), is defined as the difference in potential between two points in a conductor which, when carrying a current of one amp (A), dissipates a power of one watt

Table 1.1   SI units

Quantity	Unit	Abbreviation
Current	ampere	A
Length	metre	m
Luminous intensity	candela	cd
Mass	kilogram	kg
Temperature	Kelvin	K
Time	second	s
Matter	mol	mol

(Note that 0 K is equal to −273°C and an **interval** of 1 K is the same as an **interval** of 1°C.)

Table 1.2   Electrical quantities

Quantity	Derived unit	Abbreviation	Equivalent (in terms of fundamental units)
Capacitance	farad	F	$A\ s\ V^{-1}$
Charge	coulomb	C	$A\ s$
Energy	joule	J	$N\ m$
Force	newton	N	$kg\ m\ s^{-1}$
Frequency	hertz	Hz	s^{-1}
Illuminance	lux	lx	$lm\ m^{-2}$
Inductance	henry	H	$V\ s\ A^{-1}$
Luminous flux	lumen	lm	$cd\ sr$
Magnetic flux	weber	Wb	$V\ s$
Potential	volt	V	$W\ A^{-1}$
Power	watt	W	$J\ s^{-1}$
Resistance	ohm	Ω	$V\ A^{-1}$

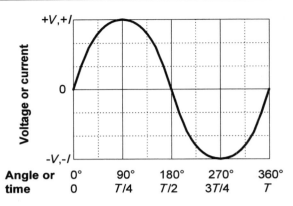

Figure 1.1   One cycle of a sine wave

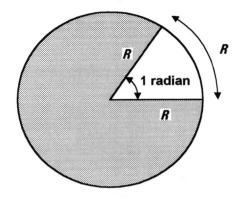

Figure 1.2   Definition of the radian

(W). Express the volt (V) in terms of joules (J) and coulombs (C).

Solution

In terms of the derived units:

$$\text{volts} = \frac{\text{watts}}{\text{amperes}} = \frac{\text{joules/seconds}}{\text{amperes}}$$

$$= \frac{\text{joules}}{\text{amperes} \times \text{seconds}} = \frac{\text{joules}}{\text{coulombs}}$$

Note that: watts = joules/seconds and coulombs = amperes × seconds
   In terms of units:

$$V = \frac{W}{A} = \frac{J/s}{A} = \frac{J}{A\,s} = \frac{J}{C}$$

Hence one volt is equivalent to one joule per coulomb.

Measuring angles

You might think it strange to be concerned with angles in electrical circuits. The reason is simply that, in analogue and a.c. circuits, signals are based on repetitive waves (often sinusoidal in shape). We can refer to a point on such a wave in one of two basic ways, either in terms of the time from the start of the cycle or in terms of the angle (a cycle starts at 0° and finishes as 360° (see Fig. 1.1). In practice, it is often more convenient to use angles rather than time, however, the two methods of measurement are interchangeable.
   In electrical circuits, angles are measured in

either degrees or radians (both of which are strictly dimensionless units). You will doubtless already be familiar with angular measure in degrees where one complete circular revolution is equivalent to an angular change of 360°. The alternative method of measuring angles, the **radian**, is defined somewhat differently. It is the angle subtended at the centre of a circle by an arc having length which is equal to the radius of the circle (see Fig. 1.2).
   It is often necessary to convert from radians to degrees, and vice versa. A complete circular revolution is equivalent to a rotation of 360° or 2π radians (note that π is approximately equal to 3.142). Thus one radian is equivalent to $360/2\pi$ degrees (or approximately 57.3°). The following rules should assist you when it is necessary to convert angles expressed in degrees to radians and vice versa.

(a) To convert from degrees to radians, divide by 57.3.
(b) To convert from radians to degrees, multiply by 57.3.

Example 1.3

Express a quarter of a cycle revolution in terms of:

(a) degrees;
(b) radians.

Solution

(a) There are 360° in one complete cycle (i.e. one revolution). Hence there are 360/4 or 90° in one quarter of a cycle.
(b) There are 2π radians in one complete cycle (i.e. one revolution). Hence there are $2\pi/4$ or $\pi/2$ radians in one quarter of a cycle.

Example 1.4

Express an angle of 215° in radians.

Solution

To convert from degrees to radians, divide by 57.3. Hence 215° is equivalent to 215/57.3 = 3.75 radians.

Example 1.5

Express an angle of 2.5 radians in degrees.

Solution

To convert from radians to degrees, multiply by 57.3. Hence 2.5 radians is equivalent to 2.5 × 57.3 = 143.25°.

Electrical units and symbols

You will find that the following units and symbols are commonly encountered in electrical circuits. It is important to get to know these units and also be able to recognize their abbreviations and symbols (see Table 1.3).

Multiples and sub-multiples

Unfortunately, many of the derived units are somewhat cumbersome for everyday use but we can make life a little easier by using a standard range of multiples and sub-multiples (see Table 1.4).

Example 1.6

An indicator lamp requires a current of 0.075 A. Express this in mA.

Solution

We can express the current in mA (rather than in A) by simply moving the decimal point three places to the right. Hence 0.075 A is the same as 75 mA.

Example 1.7

A medium-wave radio transmitter operates on a frequency of 1495 kHz. Express its frequency in MHz.

Table 1.3 Electrical units

Unit	Abbrev.	Symbol	Notes
Ampere	A	I	Unit of electric current (a current of 1 A flows in a conductor when a charge of 1 C is transported in a time interval of 1 s)
Coulomb	C	Q	Unit of electric charge or quantity of electricity
Farad	F	C	Unit of capacitance (a capacitor has a capacitance of 1 F when a charge of 1 C results in a potential difference of 1 V across its plates)
Henry	H	L	Unit of inductance (an inductor has an inductance of 1 H when an applied current changing uniformly at a rate of 1 A/s produces a potential difference of 1 V across its terminals)
Hertz	Hz	f	Unit of frequency (a signal has a frequency of 1 Hz if one complete cycle occurs in a time interval of 1 s).
Joule	J	E	Unit of energy
Ohm	ω	R	Unit of resistance
Second	s	t	Unit of time
Siemen	S	G	Unit of conductance (the reciprocal of resistance)
Tesla	T	B	Unit of magnetic flux density (a flux density of 1 T is produced when a flux of 1 Wb is present over an area of 1 square metre)
Volt	V	V	Unit of electric potential (e.m.f. or p.d.)
Watt	W	P	Unit of power (equal to 1 J of energy consumed in a time of 1 s)
Weber	Wb	Φ	Unit of magnetic flux

Table 1.4   Multiples and sub-multiples

Prefix	Abbrev.	Multiplier	
tera	T	10^{12}	(=1 000 000 000 000)
giga	G	10^{9}	(=1 000 000 000)
mega	M	10^{6}	(=1 000 000)
kilo	k	10^{3}	(=1 000)
(none)	(none)	10^{0}	(=1)
centi	c	10^{-2}	(=0.01)
milli	m	10^{-3}	(=0.001)
micro	μ	10^{-6}	(=0.000 001)
nano	n	10^{-9}	(=0.000 000 001)
pico	p	10^{-12}	(=0.000 000 000 001)

Solution

To express the frequency in MHz rather than kHz we need to move the decimal point three places to the left. Hence 1495 kHz is equivalent to 1.495 MHz.

Example 1.8

A capacitor has a value of 27 000 pF. Express this in μF.

Solution

To express the value in μF rather than pF we need to move the decimal point six places to the left. Hence 27 000 pF is equivalent to 0.027 μF (note that we have had to introduce an extra zero before the 2 and after the decimal point).

Exponent notation

Exponent notation (or **scientific notation**) is useful when dealing with either very small or very large quantities. It is well worth getting to grips with this notation as it will allow you to simplify quantities before using them in formulae.

Exponents are based on **powers of ten**. To express a number in exponent notation the number is split into two parts. The first part is usually a number in the range 0.1 to 100 while the second part is a multiplier expressed as a power of ten. For example, 251.7 can be expressed as 2.517×100, i.e. 2.517×10^{2}. It can also be expressed as 0.2517×1000, i.e. 0.2517×10^{3}. In both cases the exponent is the same as the number of noughts in the multiplier (i.e. 2 in the first case and 3 in the second case). To summarize:

$$251.7 = 2.517 \times 10^{2} = 0.2517 \times 10^{3}$$

As a further example, 0.01825 can be expressed as 1.825/100, i.e. 1.825×10^{-2}. It can also be expressed as 18.25/1000, i.e. 18.25×10^{-3}. Again, the exponent is the same as the number of noughts but the minus sign is used to denote a fractional multiplier. To summarize:

$$0.01825 = 1.825 \times 10^{-2} = 18.25 \times 10^{-3}$$

Example 1.9

A current of 7.25 mA flows in a circuit. Express this current in amperes using exponent notation.

Solution

$1 \text{ mA} = 1 \times 10^{-3} \text{ A}$ thus $7.25 \text{ mA} = 7.25 \times 10^{-3} \text{ A}$

Example 1.10

A voltage of 3.75×10^{-6} V appears at the input of an amplifier. Express this voltage in volts using exponent notation.

Solution

$1 \times 10^{-6} \text{ V} = 1 \text{ μV}$ thus $3.75 \times 10^{-6} \text{ V} = 3.75 \text{ μV}$

Multiplication and division using exponents

Exponent notation really comes into its own when values have to be multiplied or divided. When multiplying two values expressed using exponents, it is simply necessary to add the exponents. As an example:

$$(2 \times 10^{2}) \times (3 \times 10^{6}) = (2 \times 3) \times 10^{(2+6)} = 6 \times 10^{8}$$

Similarly, when dividing two values which are expressed using exponents, it is simply necessary to subtract the exponents. As an example:

$$(4 \times 10^{6}) \div (2 \times 10^{4}) = 4/2 \times 10^{(6-4)} = 2 \times 10^{2}$$

In either case it is essential to take care to express the units, multiples and sub-multiples in which you are working.

Example 1.11

A current of 3 mA flows in a resistance of 33 kΩ. Determine the voltage dropped across the resistor.

Solution

Voltage is equal to current multiplied by resistance (see page 6). Thus:

$V = I \times R = 3 \text{ mA} \times 33 \text{ k}\Omega$

Expressing this using exponent notation gives:

$V = (3 \times 10^{-3}) \times (33 \times 10^{3})$ volts

Separating the exponents gives:

$V = 3 \times 33 \times 10^{-3} \times 10^{3}$ volts

Thus $V = 99 \times 10^{(-3+3)} = 99 \times 10^{0} = 99 \times 1 = 99$ V

Example 1.12

A current of 45 µA flows in a circuit. What charge is transferred in a time interval of 20 ms?

Solution

Charge is equal to current multiplied by time (see the definition of the ampere on page 3). Thus:

$Q = it = 45 \text{ µA} \times 20 \text{ ms}$

Expressing this in exponent notation gives:

$Q = (45 \times 10^{-6}) \times (20 \times 10^{-3})$ coulomb

Separating the exponents gives:

$Q = 45 \times 20 \times 10^{-6} \times 10^{-3}$ coulomb

Thus $Q = 900 \times 10^{(-6-3)} = 900 \times 10^{(-9)} = 900$ nC

Example 1.13

A power of 300 mW is dissipated in a circuit when a voltage of 1500 V is applied. Determine the current supplied to the circuit.

Solution

Current is equal to power divided by voltage (see page 8). Thus:

$I = P/V = 300 \text{ mW}/1500 \text{ V}$ amperes

Expressing this in exponent notation gives:

$I = (300 \times 10^{-3})/(1.5 \times 10^{3})$ amperes

Separating the exponents gives:

$I = 300/1.5 \times 10^{-3}/10^{3}$ amperes

$I = 300/1.5 \times 10^{-3} \times 10^{-3}$ amperes

Thus $I = 200 \times 10^{(-3-3)} = 200 \times 10^{(-6)} = 200$ µA

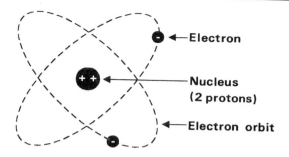

Figure 1.3 A single atom of helium (He) showing its two electrons in orbit around its nucleus

Conductors and insulators

Electric current is the name given to the flow of **electrons** (or negative charge carriers). Electrons orbit around the nucleus of atoms just as the earth orbits around the sun (see Fig. 1.3). Electrons are held in one or more **shells**, constrained to their orbital paths by virtue of a force of attraction towards the nucleus which contains an equal number of **protons** (positive charge carriers). Since like charges repel and unlike charges attract, negatively charged electrons are attracted to the positively charged nucleus. A similar principle can be demonstrated by observing the attraction between two permanent magnets; the two north poles of the magnets will repel each other, while a north and south pole will attract. In the same way, the unlike charges of the negative electron and the positive proton experience a force of mutual attraction.

The outer shell electrons of a **conductor** can be reasonably easily interchanged between adjacent atoms within the **lattice** of atoms of which the substance is composed. This makes it possible for the material to conduct electricity. Typical examples of conductors are metals such as copper, silver, iron and aluminium. By contrast, the outer shell electrons of an **insulator** are firmly bound to their parent atoms and virtually no interchange of electrons is possible. Typical examples of insulators are plastics, rubber and ceramic materials.

Voltage and resistance

The ability of an energy source (e.g. a battery) to produce a current within a conductor may be expressed in terms of **electromotive force** (e.m.f.). Whenever an e.m.f. is applied to a circuit **a potential difference** (p.d.) exists. Both e.m.f. and p.d. are

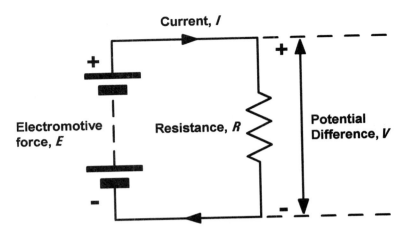

Current, _I_

Electromotive force, _E_

Resistance, _R_

Potential Difference, _V_

Figure 1.4   Simple circuit to illustrate the relationship between voltage (V), current (I) and resistance (R) (note the direction of conventional current flow from positive to negative)

measured in volts (V). In many practical circuits there is only one e.m.f. present (the battery or supply) whereas a p.d. will be developed across each component present in the circuit.

The conventional flow of current in a circuit is from the point of more positive potential to the point of greatest negative potential (note that electrons move in the *opposite* direction!). **Direct current** results from the application of a direct e.m.f. (derived from batteries or d.c. supply rails). An essential characteristic of such supplies is that the applied e.m.f. does not change its polarity (even though its value might be subject to some fluctuation).

For any conductor, the current flowing is directly proportional to the e.m.f. applied. The current flowing will also be dependent on the physical dimensions (length and cross-sectional area) and material of which the conductor is composed. The amount of current that will flow in a conductor when a given e.m.f. is applied is inversely proportional to its **resistance**. Resistance, therefore, may be thought of as an opposition to current flow; the higher the resistance the lower the current that will flow (assuming that the applied e.m.f. remains constant).

Ohm's law

Provided that temperature does not vary, the ratio of p.d. across the ends of a conductor to the current flowing in the conductor is a constant. This relationship is known as Ohm's law and it leads to the relationship:

$$V/I = \text{a constant} = R$$

where V is the potential difference (or voltage drop) in volts (V), I is the current in amperes (A), and R is the resistance in ohms (Ω) (see Fig. 1.4).

The formula may be arranged to make V, I or R the subject, as follows:

$$V = I \times R \qquad I = V/R \qquad \text{and} \qquad R = V/I$$

The triangle shown in Fig. 1.5 should help you remember these three important relationships. It is important to note that, when performing calculations of currents, voltages and resistances in practical circuits it is seldom necessary to work with an accuracy of better than ±1% simply because component tolerances are invariably somewhat greater than this. Furthermore, in calculations involving Ohm's law, it is sometimes convenient to work in units of $k\Omega$ and mA (or $M\Omega$ and μA) in which case potential differences will be expressed directly in V.

Figure 1.5   Triangle showing the relationship between V, I and R

Table 1.5 Properties of common metals

Metal	Resistivity (at 20 °C) (Ωm)	Relative conductivity (copper = 1)	Temperature coefficient of resistance (per °C)
Silver	1.626×10^{-8}	1.06	0.0041
Copper (annealed)	1.724×10^{-8}	1.00	0.0039
Copper (hard drawn)	1.777×10^{-8}	0.97	0.0039
Aluminium	2.803×10^{-8}	0.61	0.0040
Mild steel	1.38×10^{-7}	0.12	0.0045
Lead	2.14×10^{-7}	0.08	0.0040

Example 1.14

A 12 Ω resistor is connected to a 6 V battery. What current will flow in the resistor?

Solution

Here we must use $I = V/R$ (where $V = 6$ V and $R = 12$ Ω):

$I = V/R = 6$ V/12 $\Omega = 0.5$ A (or 500 mA)

Hence a current of 500 mA will flow in the resistor.

Example 1.15

A current of 100 mA flows in a 56 Ω resistor. What voltage drop (potential difference) will be developed across the resistor?

Solution

Here we must use $V = I \times R$ and ensure that we work in units of volts (V), amperes (A) and ohms (Ω).

$V = I \times R = 0.1$ A $\times$ 56 $\Omega = 5.6$ V

(Note that 100 mA is the same as 0.1 A.)

Hence a p.d. of 5.6 V will be developed across the resistor.

Example 1.16

A voltage drop of 15 V appears across a resistor in which a current of 1 mA flows. What is the value of the resistance?

Solution

$R = V/I = 15$ V/0.001 A = 15 000 Ω = 15 kΩ

Note that it is often more convenient to work in units of mA and V which will produce an answer directly in kΩ, i.e.

$R = V/I = 15$ V/1 mA = 15 kΩ

Resistance and resistivity

The resistance of a metallic conductor is directly proportional to its length and inversely proportional to its area. The resistance is also directly proportional to its **resistivity** (or **specific resistance**). Resistivity is defined as the resistance measured between the opposite faces of a cube having sides of 1 cm.

The resistance, R, of a conductor is thus given by the formula:

$R = \rho \times l/A$

where R is the resistance (Ω), ρ is the resistivity (Ω m), l is the length (m), and A is the area (m^2). Table 1.5 shows the electrical properties of various metals.

Example 1.17

A coil consists of an 8 m length of annealed copper wire having a cross-sectional area of 1 mm^2. Determine the resistance of the coil.

Solution

We will use the formula, $R = \rho l/A$.

The value of ρ for copper is 1.724×10^{-8} Ω m given in Table 1.5 which shows the properties of common metallic conductors. The length of the wire is 4 m while the area is 1 mm^2 or 1×10^{-6} m^2 (note that it is important to be consistent in using units of metres for length and square metres for area).

Hence the resistance of the coil will be given by:

$$R = \frac{1.724 \times 10^{-8} \times 8}{1 \times 10^{-6}}$$

Thus $R = 13.792 \times 10^{-2}$ or 0.137 92 Ω.

Example 1.18

A wire having a resistivity of 1.6×10^{-8} Ω m, length 20 m and cross-sectional area 1 mm² carries a current of 5 A. Determine the voltage drop between the ends of the wire.

Solution

First we must find the resistance of the wire (as in Example 1.17):

$$R = \rho l/A = \frac{1.6 \times 10^{-8} \times 20}{1 \times 10^{-6}} = 0.32 \ \Omega$$

The voltage drop can now be calculated using Ohm's law:

$$V = I \times R = 5 \text{ A} \times 0.32 \ \Omega = 1.6 \text{ V}$$

Hence a potential of 1.6 V will be dropped between the ends of the wire.

Energy and power

At first you may be a little confused about the difference between energy and power. Energy is the ability to do work while power is the rate at which work is done. In electrical circuits, energy is supplied by batteries or generators. It may also be stored in components such as capacitors and inductors. Electrical energy is converted into various other forms of energy by components such as resistors (producing heat), loudspeakers (producing sound energy) and light emitting diodes (producing light).

The unit of energy is the joule (J). Power is the rate of use of energy and it is measured in watts (W). A power of 1 W results from energy being used at the rate of 1 J per second. Thus:

$$P = E/t$$

where P is the power in watts (W), E is the energy in joules (J), and t is the time in seconds (s).

The power in a circuit is equivalent to the product of voltage and current. Hence:

$$P = I \times V$$

where P is the power in watts (W), I is the current in amperes (A), and V is the voltage in volts (V).

The formula may be arranged to make P, I or V the subject, as follows:

$$P = I \times V \qquad I = P/V \qquad \text{and} \qquad V = P/I$$

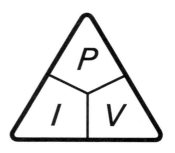

Figure 1.6   Triangle showing the relationship between P, I and V

The triangle shown in Fig. 1.6 should help readers remember these relationships.

The relationship, $P = I \times V$, may be combined with that which results from Ohm's law ($V = I \times R$) to produce two further relationships. First, substituting for V gives:

$$P = I \times (I \times R) = I^2R$$

Secondly, substituting for I gives:

$$P = (V/R) \times V = V^2/R$$

Example 1.19

A current of 1.5 A is drawn from a 3 V battery. What power is supplied? Here we must use $P = I \times V$ (where $I = 1.5$ A and $V = 3$ V):

Solution

$$P = I \times V = 1.5 \text{ A} \times 3 \text{ V} = 4.5 \text{ W}$$

Hence a power of 4.5 W is supplied.

Example 1.20

A voltage drop of 4 V appears across a resistor of 100 Ω. What power is dissipated in the resistor?

Solution

Here we use $P = V^2/R$ (where $V = 4$ V and $R = 100$ Ω):

$$P = V^2/R = (4 \text{ V} \times 4 \text{ V})/100 \ \Omega = 0.16 \text{ W}$$

Hence the resistor dissipates a power of 0.16 W (or 160 mW).

Example 1.21

A current of 20 mA flows in a 1 kΩ resistor. What power is dissipated in the resistor?

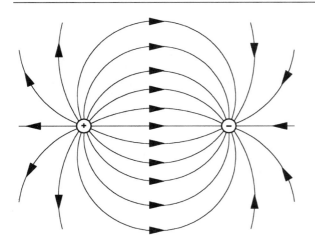

Figure 1.7 Electric fields between unlike electric charges

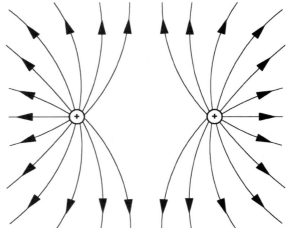

Figure 1.8 Electric fields between like electric charges

Solution

Here we use $P = I^2 \times R$ but, to make life a little easier, we will work in mA and kΩ (in which case the answer will be in mW):

$P = I^2 \times R = (20 \text{ mA} \times 20 \text{ mA}) \times 1 \text{ kΩ}$

$\quad = 400 \text{ mW}$

Thus a power of 400 mW is dissipated in the resistor.

Electrostatics

If a conductor has a deficit of electrons, it will exhibit a net positive charge. If, on the other hand, it has a surplus of electrons, it will exhibit a net positive charge. An imbalance in charge can be produced by friction (removing or depositing electrons using materials such as silk and fur, respectively) or induction (by attracting or repelling electrons using a second body which is, respectively, positively or negatively charged).

Force between charges

Coulomb's law states that, if charged bodies exist at two points, the force of attraction (if the charges are of opposite charge) or repulsion (if of like charge) will be proportional to the product of the magnitude of the charges divided by the square of their distance apart. Thus:

$$F = \frac{kQ_1Q_2}{r^2}$$

where Q_1 and Q_2 are the charges present at the two points (in coulombs), r the distance separating the two points (in metres), F is the force (in newtons), and k is a constant depending upon the medium in which the charges exist.

In vacuum or 'free space',

$$k = \frac{1}{4\pi\varepsilon_o}$$

where ε_o is the **permittivity of free space** (8.854 $\times 10^{-12}$ C/N m^2).

Combining the two previous equations gives:

$$F = \frac{Q_1Q_2}{4\pi\varepsilon_o r^2}$$

or

$$F = \frac{Q_1Q_2}{4\pi 8.854 \times 10^{-12} r^2} \quad \text{newtons}$$

Electric fields

The force exerted on a charged particle is a manifestation of the existence of an electric field. The electric field defines the direction and magnitude of a force on a charged object. The field itself is invisible to the human eye but can be drawn by constructing lines which indicate the motion of a free positive charge within the field; the number of field lines in a particular region being used to indicate the relative strength of the field at the point in question.

Figures 1.7 and 1.8 show the electric fields

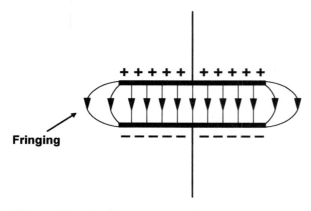

Fringing

Figure 1.9   Electric field between two charged parallel plates

between unlike and like charges while Fig. 1.9 shows the field which exists between two charged parallel plates (note the 'fringing' which occurs at the edges of the plates).

Electric field strength

The strength of an electric field (E) is proportional to the applied potential difference and inversely proportional to the distance between the two conductors. The electric field strength is given by:

$E = V/d$

where E is the electric field strength (V/m), V is the applied potential difference (V) and d is the distance (m).

Example 1.22

Two parallel conductors are separated by a distance of 25 mm. Determine the electric field strength if they are fed from a 600 V d.c. supply.

Solution

The electric field strength will be given by:

$E = V/d = 600/25 \times 10^{-3} = 24$ kV/m

Electromagnetism

When a current flows through a conductor a magnetic field is produced in the vicinity of the conductor. The magnetic field is invisible but its presence can be detected using a compass needle

(which will deflect from its normal north–south position). If two current-carrying conductors are placed in the vicinity of one another, the fields will interact with one another and the conductors will experience a force of attraction or repulsion (depending upon the relative direction of the two currents).

Force between current-carrying conductors

The mutual force which exists between two parallel current-carrying conductors will be proportional to the product of the currents in the two conductors and the length of the conductors but inversely proportional to their separation. Thus:

$$F = \frac{kI_1I_2l}{d}$$

where I_1 and I_2 are the currents in the two conductors (in amps), l is the parallel length of the conductors (in metres), d is the distance separating the two conductors (in metres), F is the force (in newtons), and k is a constant depending upon the medium in which the charges exist.

In vacuum or 'free space',

$$k = \frac{\mu_o}{2\pi}$$

where μ_o is a constant known as the **permeability of free space** (12.57×10^{-7} H/m).

Combining the two previous equations gives:

$$F = \frac{\mu_o I_1 I_2 l}{2\pi d}$$

or

$$F = \frac{4\pi \times 10^{-7} I_1 I_2 l}{2\pi d} \quad \text{newtons}$$

or

$$F = \frac{2 \times 10^{-7} I_1 I_2 l}{d} \quad \text{newtons}$$

Magnetic fields

The field surrounding a straight current-carrying conductor is shown in Fig. 1.10. The magnetic field defines the direction of motion of a free north pole within the field. In the case of Fig. 1.10, the lines

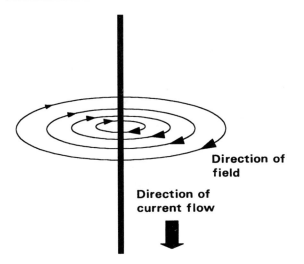

Figure 1.10 Magnetic field surrounding a straight current-carrying conductor

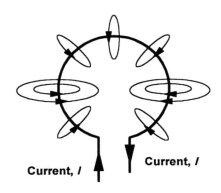

Figure 1.11 Forming a conductor into a loop increases the strength of the magnetic field in the centre of the loop

of flux are concentric and the direction of the field (determined by the direction of current flow) is given by the right-hand screw rule.

Magnetic field strength

The strength of a magnetic field is a measure of the density of the flux at any particular point. In the case of Fig. 1.10, the field strength will be proportional to the applied current and inversely proportional to the perpendicular distance from the conductor. Thus

$$B = \frac{kI}{d}$$

where B is the magnetic flux density (in tesla), I is the current (in amperes), d is the distance from the conductor (in metres), and k is a constant. Assuming that the medium is vacuum or 'free space', the density of the magnetic flux will be given by:

$$B = \frac{\mu_o I}{2\pi d}$$

where B is the flux density (in tesla), μ_o is the permeability of 'free space' ($4\pi \times 10^{-7}$ or 12.57×10^{-7}), I is the current (in amperes), and d is the distance from the centre of the conductor (in metres).

The flux density is also equal to the total flux divided by the area of the field. Thus:

$$B = \Phi/A$$

where Φ is the flux (in webers) and A is the area of the field (in square metres).

In order to increase the strength of the field, a conductor may be shaped into a loop (Fig. 1.11) or coiled to form an solenoid (Fig. 1.12). Note, in the latter case, how the field pattern is exactly the same as that which surrounds a bar magnet.

Example 1.23

Determine the flux density produced at a distance of 50 mm from a straight wire carrying a current of 20 A.

Solution

Applying the formula $B = \mu_o I / 2\pi d$ gives:

$$B = \frac{12.57 \times 10^{-7} \times 20}{2 \times 3.142 \times 50 \times 10^{-3}} = \frac{251.40 \times 10^{-7}}{314.20 \times 10^{-3}}$$

$$= 0.8 \times 10^{-4} \quad \text{tesla}$$

thus $B = 800 \times 10^{-6}$ T or $B = 80$ μT.

Example 1.24

A flux density of 2.5 mT is developed in free space over an area of 20 cm². Determine the total flux.

Solution

Re-arranging the formula $B = \Phi/A$ to make Φ the subject gives $\Phi = BA$ thus $\Phi = (2.5 \times 10^{-3}) \times (20 \times 10^{-4}) = 50 \times 10^{-7}$ Wb = 5 μWb.

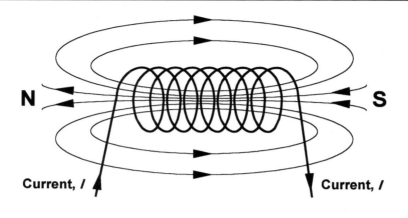

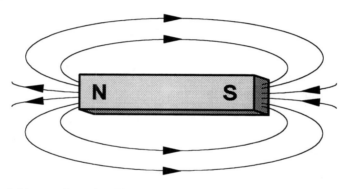

Figure 1.12   Magnetic field around a solenoid

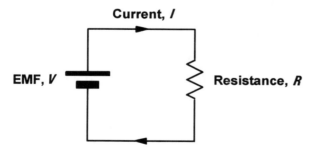

Figure 1.13   An electric circuit

Magnetic circuits

Materials such as iron and steel possess considerably enhanced magnetic properties. Hence they are employed in applications where it is necessary to increase the flux density produced by an electric current. In effect, magnetic materials allow us to channel the electric flux into a 'magnetic circuit', as shown in Fig. 1.14.

In the circuit of Fig. 1.14 the **reluctance** of the magnetic core is analogous to the resistance present in the electric circuit shown in Fig. 1.13. We can make the following comparisons between the two types of circuit (see Table 1.6):

In practice, not all of the magnetic flux produced in a magnetic circuit will be concentrated within the core and some 'leakage flux' will appear in the surrounding free space (as shown in Fig. 1.15). Similarly, if a gap appears within the magnetic circuit, the flux will tend to spread out as shown in Fig. 1.16. This effect is known as 'fringing'.

Reluctance and permeability

The reluctance of a magnetic path is directly proportional to its length and inversely proportional to its area. The reluctance is also inversely proportional

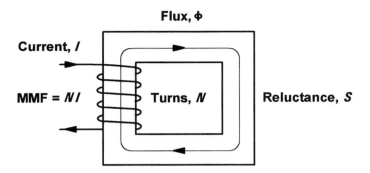

Figure 1.14   A magnetic circuit

Table 1.6   Comparison of electric and magnetic circuits

Electric circuit (Fig. 1.13)	*Magnetic circuit* (Fig. 1.14)
Electromotive force, e.m.f. = V	Magnetomotive force, m.m.f. = $N \times I$
Resistance = R	Reluctance = S
Current = I	Flux = Φ
e.m.f. = current $\times$ resistance	m.m.f. = flux $\times$ reluctance
$V = IR$	$NI = S\Phi$

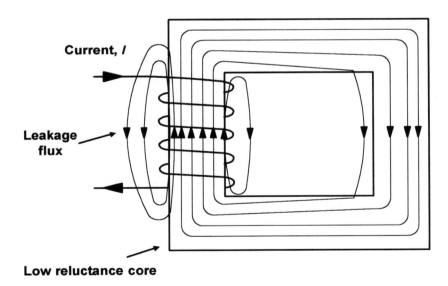

Figure 1.15   Leakage flux in a magnetic circuit

to the **absolute permeability** of the magnetic material. Thus

$$S = \frac{l}{\mu A}$$

where S is the reluctance of the magnetic path, l is the length of the path (in metres), A is the cross-sectional area of the path (in square metres), and μ is the absolute permeability of the magnetic material.

Now the absolute permeability of a magnetic material is the product of the permeability of free space (μ_o) and the **relative permeability** of the magnetic medium (μ_r). Thus

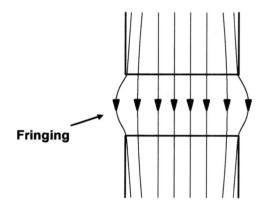

Figure 1.16   Fringing of the magnetic flux in an air gap

$$\mu = \mu_o \times \mu_r \text{ and } S = \frac{l}{\mu_o\mu_r A}$$

The permeability of a magnetic medium is a measure of its ability to support magnetic flux and it is equal to the ratio of flux density (B) to **magnetizing force** (H). Thus

$$\mu = \frac{B}{H}$$

where B is the flux density (in tesla) and H is the magnetizing force (in ampere/metre).

The magnetizing force (H) is proportional to the product of the number of turns and current but inversely proportional to the length of the magnetic path. Thus:

$$H = \frac{N \times I}{l}$$

where H is the magnetizing force (in ampere/metre), N is the number of turns, I is the current (in amperes), and l is the length of the magnetic path (in metres).

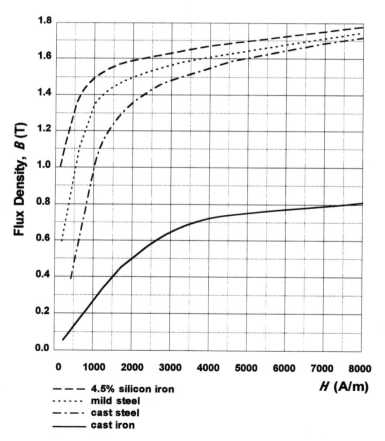

Figure 1.17 B–H curves for four magnetic materials

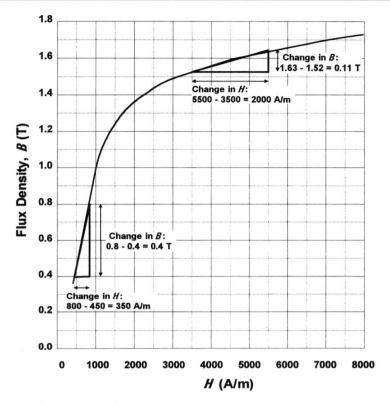

Figure 1.18 *B–H* curve for a sample of cast steel

B–H curves

Figure 1.17 shows four typical *B–H* (flux density plotted against permeability) curves for some common magnetic materials. It should be noted that each of these curves eventually flattens off due to magnetic saturation and that the slope of the curve (indicating the value of μ corresponding to a particular value of *H*) falls as the magnetizing force increases. This is important since it dictates the acceptable working range for a particular magnetic material when used in a magnetic circuit.

Example 1.25

Estimate the relative permeability of cast steel (see Fig. 1.18) at (a) a flux density of 0.6 T and (b) a flux density of 1.6 T.

Solution

From Fig. 1.18, the slope of the graph at any point gives the value of μ at that point. The slope can be found by constructing a tangent at the point in question and finding the ratio of vertical change to horizontal change.

(a) The slope of the graph at 0.6 T is $0.3/500 = 0.6 \times 10^{-3}$
Since $\mu = \mu_0\mu_r$, $\mu_r = \mu/\mu_0 = 0.6 \times 10^{-3}/12.57 \times 10^{-7}$,
thus $\mu_r = 477$.

(a) The slope of the graph at 1.6 T is $0.09/1500 = 0.06 \times 10^{-3}$
Since $\mu = \mu_0\mu_r$, $\mu_r = \mu/\mu_0 = 0.06 \times 10^{-3}/12.57 \times 10^{-7}$,
thus $\mu_r = 47.7$.

NB: This example clearly shows the effect of saturation on the permeability of a magnetic material!

Example 1.26

A coil of 800 turns is wound on a closed mild steel core having a length 600 mm and cross-sectional area 500 mm². Determine the current required to establish a flux of 0.8 mWb in the core.

Solution

$B = \Phi/A = 0.8 \times 10^{-3}/500 \times 10^{-6} = 1.6$ T

From Fig. 1.17, a flux density of 1.6 T will occur in mild steel when $H = 3500$ A/m. The current can now be determined by re-arranging $H = (N \times I)/l$

$$I = \frac{H \times l}{N} = \frac{3500 \times 0.6}{800} = 2.625 \text{ A}$$

Circuit symbols introduced in this chapter

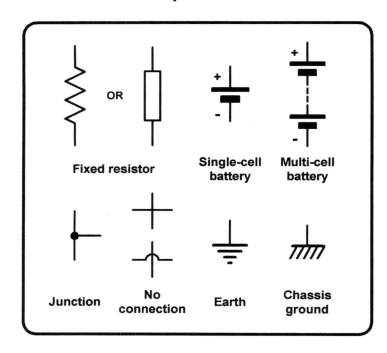

Figure 1.19

Important formulae introduced in this chapter

Voltage, current and resistance (Ohm's law): (page 6)

$V = IR$

Resistance and resistivity: (page 7)

$R = \rho l/A$

Charge, current and time: (page 5)

$Q = It$

Power, current and voltage: (page 8)

$P = IV$

Power, voltage and resistance: (page 8)

$P = V^2/R$

Power, current and resistance: (page 8)

$P = I^2R$

Reluctance and permeability: (page 13)

$$S = \frac{l}{\mu A}$$

Flux and flux density:
(page 11)

$$B = \Phi/A$$

Current and magnetic field intensity:
(page 14)

$$H = \frac{NI}{l}$$

Flux, current and reluctance:
(page 13)

$$NI = S\Phi$$

Problems

1.1 Which of the following are not fundamental units; amperes, metres, coulombs, joules, hertz, kilogram?

1.2 A commonly used unit of consumer energy is the kilowatt hour (kWh). Express this in joules (J).

1.3 Express an angle of 30° in radians.

1.4 Express an angle of 0.2 radians in degrees.

1.5 A resistor has a value of 39 570 Ω. Express this in kilohms (kΩ).

1.6 An inductor has a value of 680 mH. Express this in henries (H).

1.7 A capacitor has a value of 0.002 45 μF. Express this in nanofarads (nF).

1.8 A current of 190 μA is applied to a circuit. Express this in milliamperes (mA).

1.9 A signal of 0.475 mV appears at the input of an amplifier. Express this in volts using exponent notation.

1.10 A cable has an insulation resistance of 16.5 MΩ. Express this resistance in ohms using exponent notation.

1.11 Perform the following arithmetic using exponents:

    (a) $(1.2 \times 10^3) \times (4 \times 10^3)$
    (b) $(3.6 \times 10^6) \times (2 \times 10^{-3})$
    (c) $(4.8 \times 10^9) \div (1.2 \times 10^6)$
    (d) $(9.9 \times 10^{-6}) \div (19.8 \times 10^{-3})$

1.12 Which one of the following metals is the best conductor of electricity: aluminium, copper, silver, or mild steel? Why?

1.13 A resistor of 270 Ω is connected across a 9 V d.c. supply. What current will flow?

1.14 A current of 56 μA flows in a 120 kΩ resistor. What voltage drop will appear across the resistor?

1.15 A voltage drop of 13.2 V appears across a resistor when a current of 4 mA flows in it. What is the value of the resistor?

1.16 A power supply is rated at 15 V, 1 A. What value of load resistor would be required to test the power supply at its full rated output?

1.17 A wirewound resistor is made from a 4 m length of aluminium wire ($\rho = 2.18 \times 10^{-8}$ Ω m) having a cross-sectional area of 0.2 mm². Determine the value of resistance.

1.18 A current of 25 mA flows in a 47 Ω resistor. What power is dissipated in the resistor?

1.19 A 9 V battery supplies a circuit with a current of 75 mA. What power is consumed by the circuit?

1.20 A resistor of 150 Ω is rated at 0.5 W. What is the maximum current that can be applied to the resistor without exceeding its rating?

1.21 Determine the electric field strength that appears in the space between two parallel plates separated by an air gap of 4 mm if a potential of 2.5 kV exists between them.

1.22 Determine the current that must be applied to a straight wire conductor in order to produce a flux density of 200 μT at a distance of 12 mm in free space.

1.23 A flux density of 1.2 mT is developed in free space over an area of 50 cm². Determine the total flux present.

1.24 A ferrite rod has a length of 250 mm and a diameter of 10 mm. Determine the reluctance if the rod has a relative permeability of 2500.

1.25 A coil of 400 turns is wound on a closed mild steel core having a length 400 mm and cross-sectional area 480 mm². Determine the current required to establish a flux of 0.6 mWb in the core.

(The answers to these problems appear on page 202.)

2

Passive components

This chapter introduces several of the most common types of electronic component, including resistors, capacitors and inductors. These are often referred to as **passive components** as they cannot, by themselves, generate voltage or current. An understanding of the characteristics and application of passive components is an essential prerequisite to understanding the operation of the circuits used in amplifiers, oscillators, filters and power supplies.

Resistors

The notion of resistance as opposition to current was discussed in the previous chapter. Conventional forms of resistor obey a straight line law when voltage is plotted against current (see Fig. 2.1) and this allows us to use resistors as a means of converting current into a corresponding voltage drop, and vice versa (note that doubling the applied current will produce double the voltage drop, and so on). Therefore resistors provide us with a means of controlling the currents and voltages present in electronic circuits. They can also act as **loads** to simulate the presence of a circuit during testing (e.g. a suitably rated resistor can be used to replace

a loudspeaker when an audio amplifier is being tested).

The specifications for a resistor usually include the value of resistance (expressed in ohms (Ω), kilohms (kΩ) or megohms (MΩ)), the accuracy or tolerance (quoted as the maximum permissible percentage deviation from the marked value), and the power rating (which must be equal to, or greater than, the maximum expected power dissipation).

Other practical considerations when selecting resistors for use in a particular application include temperature coefficient, noise performance, stability and ambient temperature range. Table 2.1 summarizes the properties of five of the most common types of resistor. Figures 2.2 and 2.3 show the construction of typical carbon rod (now obsolete) and carbon film resistors.

Preferred values

The value marked on the body of a resistor is not its *exact* resistance. Some minor variation in resistance value is inevitable due to production tolerance. For example, a resistor marked 100 Ω and produced within a tolerance of ±10% will have a value which falls within the range 90 Ω to 110 Ω. If a particular circuit requires a resistance of, for example, 105 Ω, a ±10% tolerance resistor of 100 Ω will be perfectly adequate. If, however, we need a component with a value of 101 Ω, then it would be necessary to obtain a 100 Ω resistor with a tolerance of ±1%.

Resistors are available in several series of fixed decade values, the number of values provided with each series being governed by the tolerance involved. In order to cover the full range of resistance values using resistors having a ±20% tolerance it will be necessary to provide six basic values (known as the **E6 series**). More values will be required in the series which offers a tolerance of ±10% and consequently the **E12 series** provides twelve basic values. The **E24 series** for resistors of ±5% tolerance provides no fewer than 24 basic values

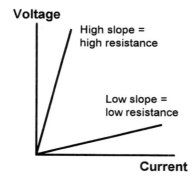

Figure 2.1 Voltage plotted against current for two different values of resistor (note that the slope of the graph is proportional to the value of resistance)

Table 2.1   Characteristics of common types of resistor

Parameter	Resistor type				
	Carbon film	Metal film	Metal oxide	Ceramic wirewound	Vitreous wirewound
Resistance range (Ω)	10 to 10 M	1 to 1 M	10 to 1 M	0.47 to 22 k	0.1 to 22 k
Typical tolerance (%)	±5	±1	±2	±5	±5
Power rating (W)	0.25 to 2	0.125 to 0.5	0.25 to 0.5	4 to 17	2 to 4
Temperature coefficient (ppm/°C)	−250	+50 to +100	+250	+250	+75
Stability	Fair	Excellent	Excellent	Good	Good
Noise performance	Fair	Excellent	Excellent	n/a	n/a
Ambient temperature range (°C)	−45 to +125	−55 to +125	−55 to +155	−55 to +200	−55 to +200
Typical applications	General purpose	Amplifiers, test equipment, etc. requiring low-noise high-tolerance components		Power supplies, loads, high-power circuits	

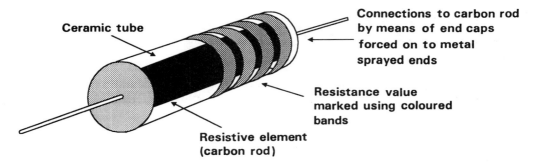

Figure 2.2   Construction of a carbon rod resistor

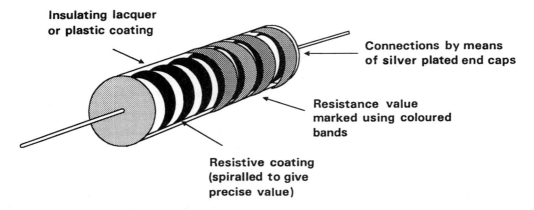

Figure 2.3   Construction of a carbon film resistor

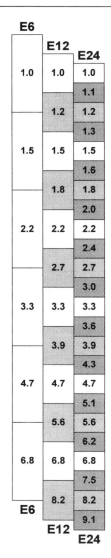

Figure 2.4   The E6, E12 and E24 series

and, as with the E6 and E12 series, decade multiples (i.e. ×1, ×10, ×100, ×1k, ×10k, ×100k and ×1M) of the basic series. Figure 2.4 shows the relationship between the E6, E12 and E24 series.

Power ratings

Resistor power ratings are related to operating temperatures and resistors should be derated at high temperatures. Where reliability is important resistors should be operated at well below their nominal maximum power dissipation.

Example 2.1

A resistor has a marked value of 220 Ω. Determine the tolerance of the resistor if it has a measured value of 207 Ω.

Solution

The difference between the marked and measured values of resistance (the error) is (220 Ω – 207 Ω) = 13 Ω. The tolerance is given by:

$$\text{Tolerance} = \frac{\text{error}}{\text{marked value}} \times 100\%$$

The tolerance is thus $13/220 \times 100 = 5.9\%$.

Example 2.2

A 9 V power supply is to be tested with a 39 Ω load resistor. If the resistor has a tolerance of 10% determine:

(a) the nominal current taken from the supply;
(b) the maximum and minimum values of supply current at either end of the tolerance range for the resistor.

Solution

(a) If a resistor of **exactly** 39 Ω is used the current will be:

$$I = V/R = 9 \text{ V}/39 \ \Omega = 231 \text{ mA}$$

(b) The lowest value of resistance would be (39 Ω – 3.9 Ω) = 35.1 Ω. In which case the current would be:

$$I = V/R = 9 \text{ V}/35.1 \ \Omega = 256.4 \text{ mA}$$

At the other extreme, the highest value of resistance would be (39 Ω + 3.9 Ω) = 42.9 Ω. In this case the current would be:

$$I = V/R = 9 \text{ V}/42.9 \ \Omega = 209.8 \text{ mA}$$

Example 2.3

A current of 100 mA ($\pm20\%$) is to be drawn from a 28 V d.c. supply. What value and type of resistor should be used in this application?

Solution

The value of resistance required must first be calculated using Ohm's law:

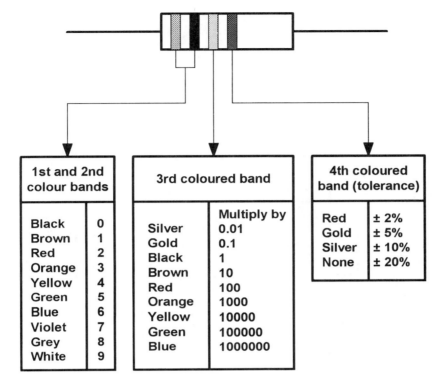

Figure 2.5   Four band resistor colour code

$R = V/I = 28$ V$/100$ mA $= 280\ \Omega$

The nearest preferred value from the E12 series is 270 Ω (which will actually produce a current of 103.7 mA (i.e. within ±4% of the desired value). If a resistor of ±10% tolerance is used, current will be within the range 94 mA to 115 mA (well within the ±20% accuracy specified). The power dissipated in the resistor (calculated using $P = I \times V$) will be 2.9 W and thus a component rated at 3 W (or more) will be required. This would normally be a vitreous enamel coated wirewound resistor (see Table 2.1).

Resistor markings

Carbon and metal oxide resistors are normally marked with colour codes which indicate their value and tolerance. Two methods of colour coding are in common use; one involves four coloured bands (see Fig. 2.5) while the other uses five colour bands (see Fig. 2.6).

Example 2.4

A resistor is marked with the following coloured stripes: brown, black, red, silver. What is its value and tolerance?

Solution

See Fig. 2.7.

Example 2.5

A resistor is marked with the following coloured stripes: red, violet, orange, gold. What is its value and tolerance?

Solution

See Fig. 2.8.

Example 2.6

A resistor is marked with the following coloured stripes: green, blue, black, gold. What is its value and tolerance?

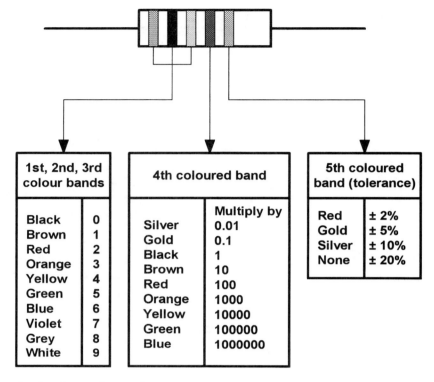

Figure 2.6   Five band resistor colour code

Solution

See Fig. 2.9.

Example 2.7

A resistor is marked with the following coloured stripes: red, green, black, black, brown. What is its value and tolerance?

Solution

See Fig. 2.10.

BS 1852 coding

Some types of resistor have markings based on a system of coding defined in BS 1852. This system involves marking the position of the decimal point with a letter to indicate the multiplier concerned as shown in Table 2.2. A further letter is then appended to indicate the tolerance as shown in Table 2.3.

Example 2.8

A resistor is marked coded with the legend 4R7K. What is its value and tolerance?

Solution

4.7 Ω ±10%

Example 2.9

A resistor is marked coded with the legend 330RG. What is its value and tolerance?

Solution

330 Ω ±2%

Example 2.10

A resistor is marked coded with the legend R22M. What is its value and tolerance?

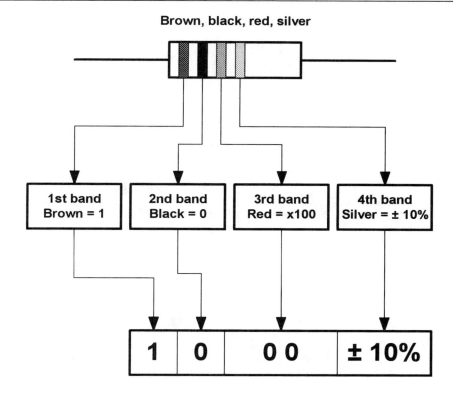

Figure 2.7

Solution

0.22 Ω ±20%

Series and parallel combinations of resistors

In order to obtain a particular value of resistance, fixed resistors may be arranged in either series or parallel as shown in Figs 2.11 and 2.12.

The effective resistance of each of the series circuits shown in Fig. 2.11 is simply equal to the sum of the individual resistances. Hence, for Fig. 2.11(a)

$$R = R_1 + R_2$$

while for Fig. 2.11(b)

$$R = R_1 + R_2 + R_3$$

Turning to the parallel resistors shown in Fig. 2.12, the reciprocal of the effective resistance of each circuit is equal to the sum of the reciprocals of the individual resistances. Hence, for Fig. 2.12(a)

$$\frac{1}{R} = \frac{1}{R_1} + \frac{1}{R_2}$$

while for Fig. 2.12(b)

$$\frac{1}{R} = \frac{1}{R_1} + \frac{1}{R_2} + \frac{1}{R_3}$$

In the former case, the formula can be more conveniently re-arranged as follows:

$$R = \frac{R_1 \times R_2}{R_1 + R_2}$$

(You can remember this as the *product* of the two resistance values *divided by* the *sum* of the two resistance values.)

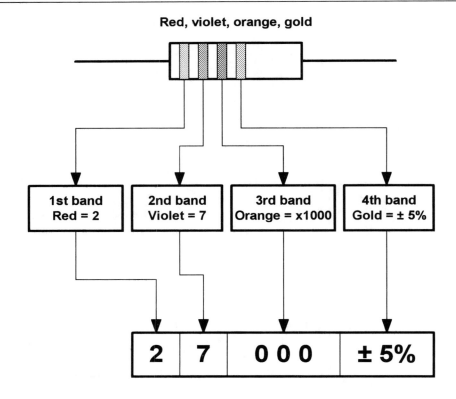

Figure 2.8

Example 2.11

Resistors of 22 Ω, 47 Ω and 33 Ω are connected (a) in series and (b) in parallel. Determine the effective resistance in each case.

Solution

(a) In the series circuit $R = R_1 + R_2 + R_3$, thus

$$R = 22\ \Omega + 47\ \Omega + 33\ \Omega = 102\ \Omega$$

(b) In the parallel circuit:

$$\frac{1}{R} = \frac{1}{R_1} + \frac{1}{R_2} + \frac{1}{R_3}$$

thus

$$\frac{1}{R} = \frac{1}{22\ \Omega} + \frac{1}{47\ \Omega} + \frac{1}{33\ \Omega}$$

or

$$\frac{1}{R} = 0.045 + 0.021 + 0.03$$

thus

$$R = \frac{1}{0.096} = 10.42\ \Omega$$

Example 2.12

Determine the effective resistance of the circuit shown in Fig. 2.13.

Solution

The circuit can be progressively simplified as shown in Fig. 2.14. The stages in this simplification are:

(a) R_3 and R_4 are in series and they can be replaced by a single resistance (R_A) of (12 Ω + 27 Ω) = 39 Ω.

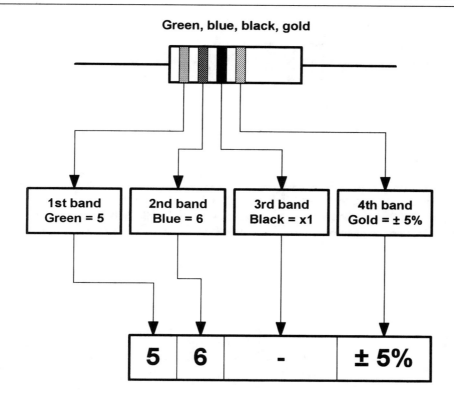

Green, blue, black, gold

1st band Green = 5	2nd band Blue = 6	3rd band Black = x1	4th band Gold = ± 5%

5	6	-	± 5%

56 ohm ± 5%

Figure 2.9

(b) R_A appears in parallel with R_2. These two resistors can be replaced by a single resistance (R_B) of (39 Ω × 47 Ω)/(39 Ω + 47 Ω) = 21.3 Ω.

(c) R_B appears in series with R_1. These two resistors can be replaced by a single resistance (R) of (21.3 Ω + 4.7 Ω) = 26 Ω.

Example 2.13

A resistance of 50 Ω 2 W is required. What parallel combination of preferred value resistors will satisfy this requirement? What power rating should each resistor have?

Solution

Two 100 Ω resistors may be wired in parallel to provide a resistance of 50 Ω as shown below:

$$R = \frac{R_1 \times R_2}{R_1 + R_2} = \frac{100 \times 100}{100 + 100} = \frac{10\,000}{200} = 50\ \Omega$$

Since the resistors are identical, the applied power will be shared equally between them. Hence each resistor should have a power rating of 1 W.

Resistance and temperature

Figure 2.15 shows how the resistance of a metal conductor (e.g. copper) varies with temperature. Since the resistance of the material increases with temperature, this characteristic is said to exhibit a **positive temperature coefficient (PTC)**. Not all materials have a PTC characteristic. The resistance of a carbon conductor falls with temperature and it is therefore said to exhibit a **negative temperature coefficient (NTC)**.

The resistance of a conductor at a temperature, t, is given by the equation:

$$R_t = R_0(1 + \alpha t + \beta t^2 + \gamma t^3 \ldots)$$

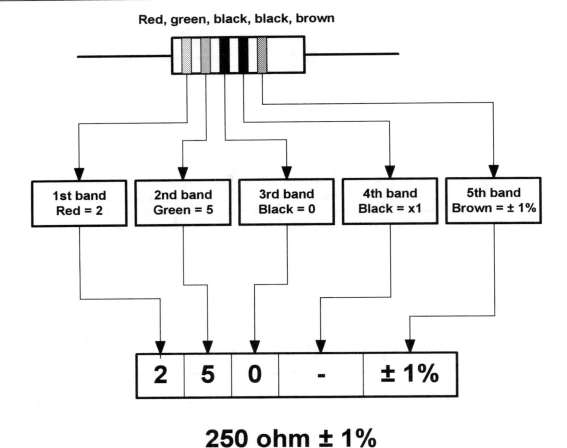

Red, green, black, black, brown

| 1st band
Red = 2 | 2nd band
Green = 5 | 3rd band
Black = 0 | 4th band
Black = x1 | 5th band
Brown = ± 1% |

| 2 | 5 | 0 | - | ± 1% |

250 ohm ± 1%

Figure 2.10

Table 2.2 Resistor multiplier markings

Letter	Multiplier
R	1
K	1 000
M	1 000 000

Table 2.3 Resistor tolerance markings

Letter	Tolerance
F	±1%
G	±2%
J	±5%
K	±10%
M	±20%

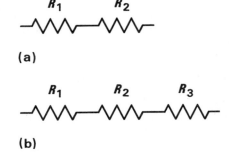

(a)

(b)

Figure 2.11 Resistors in series: (a) two resistors in series (b) three resistors in series

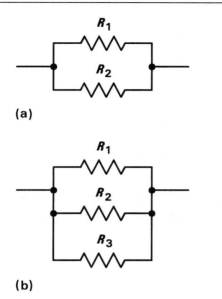

(a)

(b)

Figure 2.12   Resistors in parallel: (a) two resistors in parallel (b) three resistors in parallel

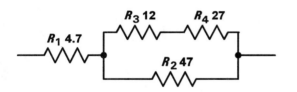

Figure 2.13   Circuit for Example 2.12

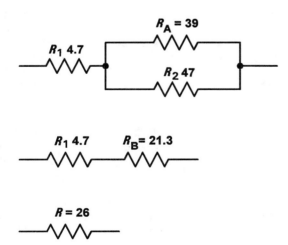

Figure 2.14   Stages in simplifying the circuit of Fig. 2.13

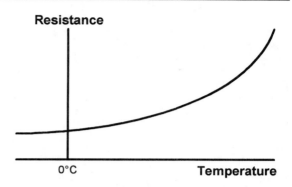

Figure 2.15   Variation of resistance with temperature for a metal conductor

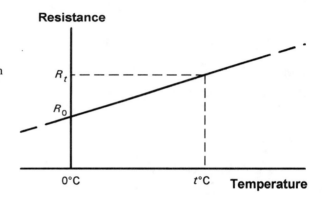

Figure 2.16   Straight line approximation of Fig. 2.15

where α, β, γ, etc. are constants and R_0 is the temperature at 0°C.

The coefficients, β, γ, etc. are quite small and since we are normally only dealing with a relatively restricted temperature range (e.g. 0°C to 100°C) we can usually approximate the characteristic shown in Fig. 2.15 to the straight line law shown in Fig. 2.16. In this case, the equation simplifies to:

$$R_t = R_0(1 + \alpha t)$$

where α is known as the **temperature coefficient** of resistance. Table 2.4 shows some typical values for α (note that α is expressed in $\Omega/\Omega/°C$ or just /°C).

Example 2.14

A resistor has a temperature coefficient of 0.001/ °C. If the resistor has a resistance of 1.5 kΩ at 0°C, determine its resistance at 80°C.

Table 2.4 Temperature coefficient of resistance

Material	Temperature coefficient of resistance, α (/°C)
Platinum	+0.0034
Silver	+0.0038
Copper	+0.0043
Iron	+0.0065
Carbon	−0.0005

Solution

Now

$R_t = R_0(1 + \alpha t)$ thus $R_t = 1.5 \text{ k}\Omega \times (1 + (0.001 \times 80))$

Hence

$R_{80} = 1.5 \times (1 + 0.08) = 1.5 \times 1.08 = 1.62 \text{ k}\Omega$

Example 2.15

A resistor has a temperature coefficient of 0.0005/°C. If the resistor has a resistance of 680 Ω at 20°C, what will its resistance be at 90°C?

Solution

First we must find the resistance at 0°C. Rearranging the formula for R_t gives:

$$R_0 = \frac{R_t}{1 + \alpha t} = \frac{680}{1 + (0.0005 \times 20)} = \frac{680}{1 + 0.01}$$

$$= \frac{680}{1.01} = 673.3 \ \Omega$$

Now

$R_t = R_0(1 + \alpha t)$ thus $R_{90} = 680 \times (1 + (0.0005 \times 90))$

Hence

$R_{90} = 680 \times (1 + (0.045)) = 680 \times 1.045$
$= 650.7 \ \Omega$

Example 2.16

A resistor has a resistance of 40 Ω at 0°C and 44 Ω at 100°C. Determine the resistor's temperature coefficient.

Solution

First we need to make α the subject of the formula $R_t = R_0(1 + \alpha t)$:
Now

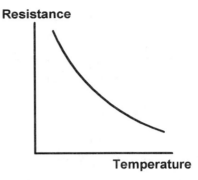

Figure 2.17 Negative temperature coefficient (NTC) thermistor characteristic

$$1 + \alpha t = \frac{R_t}{R_0} \quad \text{thus} \quad \alpha t = \frac{R_t}{R_0} - 1$$

Hence

$$\alpha = \frac{1}{t}\left(\frac{R_t}{R_0} - 1\right) = \frac{1}{100}\left(\frac{44}{40} - 1\right)$$

Therefore

$$\alpha = \frac{1}{100}(1.1 - 1) = \frac{1}{100} \times 0.1 = 0.001/°C$$

Thermistors

With conventional resistors we would normally require resistance to remain the same over a wide range of temperatures (i.e. α should be zero). On the other hand, there are applications in which we could use the effect of varying resistance to detect a temperature change. Components that allow us to do this are known as **thermistors**. The resistance of a thermistor changes markedly with temperature and these components are widely used in temperature sensing and temperature compensating applications. Two basic types of thermistor are available, NTC and PTC.

Typical NTC thermistors have resistances which vary from a few hundred (or thousand) ohms at 25°C to a few tens (or hundreds) of ohms at 100°C (see Fig. 2.17). PTC thermistors, on the other hand, usually have a resistance–temperature characteristic which remains substantially flat (typically at around 100 Ω) over the range 0°C to around 75°C. Above this, and at a critical temperature (usually in the range 80°C to 120°C) their resistance rises very rapidly to values of up to, and beyond, 10 kΩ (see Fig. 2.18).

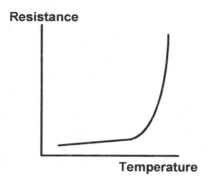

Figure 2.18   Positive temperature coefficient (PTC) thermistor characteristic

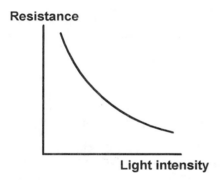

Figure 2.19   Light dependent resistor (LDR) characteristic

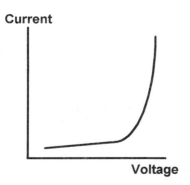

Figure 2.20   Current plotted against voltage for a voltage dependent resistor (VDR) (note that the slope of the graph is inversely proportional to the value of resistance)

A typical application of PTC thermistors is over-current protection. Provided the current passing through the thermistor remains below the threshold current, the effects of self-heating will remain negligible and the resistance of the thermistor will remain low (i.e. approximately the same as the resistance quoted at 25°C). Under fault conditions, the current exceeds the threshold value by a considerable margin and the thermistor starts to self-heat. The resistance then increases rapidly and, as a consequence, the current falls to the rest value. Typical values of threshold and rest currents are 200 mA and 8 mA, respectively, for a device which exhibits a nominal resistance of 25 Ω at 25°C.

Light dependent resistors

Light dependent resistors (LDR) use a semiconductor material (i.e. a material that is neither a conductor nor an insulator) whose electrical characteristics vary according to the amount of incident light. The two semiconductor materials used for the manufacture of LDRs are cadmium sulphide (CdS) and cadmium selenide (CdSe). These materials are most sensitive to light in the visible spectrum, peaking at about 0.6 μm for CdS and 0.75 μm for CdSe. A typical CdS LDR exhibits a resistance of around 1 MΩ in complete darkness and less than 1 kΩ when placed under a bright light source (see Fig. 2.19).

Voltage dependent resistors

The resistance of a voltage dependent resistor (VDR) falls very rapidly when the voltage across it exceeds a nominal value in either direction (see Fig. 2.20). In normal operation, the current flowing in a VDR is negligible, however, when the resistance falls, the current will become appreciable and a significant amount of energy will be absorbed.

VDRs are used as a means of 'clamping' the voltage in a circuit to a pre-determined level. When connected across the supply rails to a circuit (either AC or DC) they are able to offer a measure of protection against supply voltage surges.

Variable resistors

Variable resistors are available in several including those which use carbon tracks and those which use a wirewound resistance element. In either case, a moving slider makes contact with the resistance

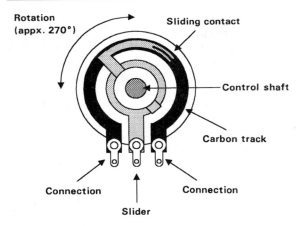

Figure 2.21 Construction of a carbon track rotary potentiometer

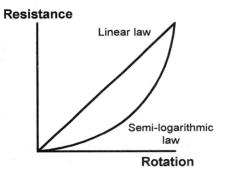

Figure 2.22 Characteristics of linear and semi-logarithmic potentiometers

element (see Fig. 2.21). Most variable resistors have three (rather than two) terminals and as such are more correctly known as potentiometers.

Carbon potentiometers are available with linear or semi-logarithmic law tracks (see Fig. 2.22) and in rotary or slider formats. Ganged controls, in which several potentiometers are linked together by a common control shaft, are also available.

You will also encounter various forms of preset resistors that are used to make occasional adjustments (e.g. for calibration). Various forms of preset resistor are commonly used including open carbon track skeleton presets and fully encapsulated carbon and multi-turn cermet types.

Capacitors

A capacitor is a device for storing electric charge. In effect, it is a reservoir into which charge can be deposited and then later extracted. Typical applications include reservoir and smoothing capacitors for use in power supplies, coupling a.c. signals between the stages of amplifiers, and decoupling supply rails (i.e. effectively grounding the supply rails as far as a.c. signals are concerned).

A capacitor need consist of nothing more than two parallel metal plates as shown in Fig. 2.23. If the switch is left open, no charge will appear on the plates and in this condition there will be no electric field in the space between the plates nor any charge stored in the capacitor.

Take a look at the circuit shown in Fig. 2.24(a).

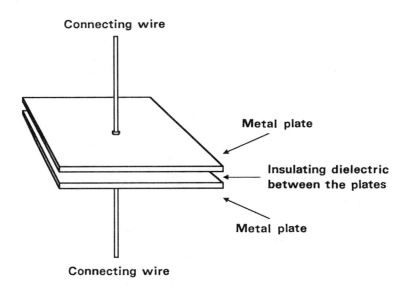

Figure 2.23 Basic parallel plate capacitor

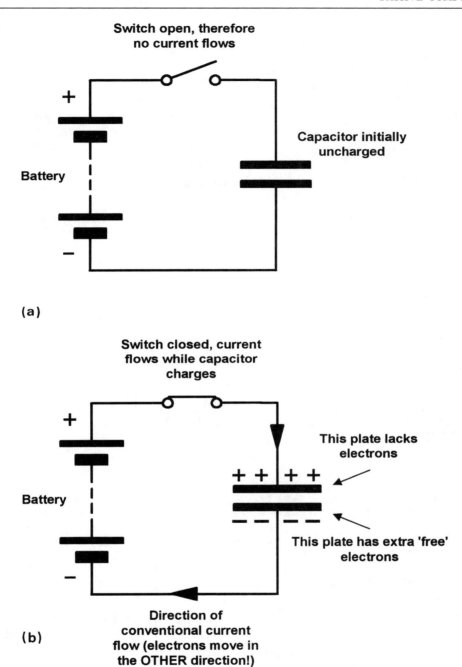

Figure 2.24 Effect of applying a voltage to a capacitor: (a) initial state (no charge present); (b) charge rapidly builds up when voltage is applied; (c) charge remains when voltage is removed

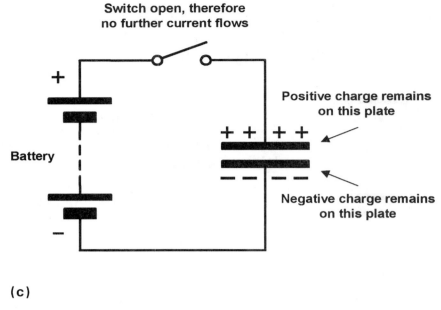

(c)

Figure 2.24   (Cont.)

With the switch left open, no current will flow and no charge will be present in the capacitor. When the switch is closed (see Fig. 2.24(b)), electrons will be attracted from the positive plate to the positive terminal of the battery. At the same time, a similar number of electrons will move from the negative terminal of the battery to the negative plate. This sudden movement of electrons will manifest itself in a momentary surge of current (conventional current will flow from the positive terminal of the battery towards the positive terminal of the capacitor).

Eventually, enough electrons will have moved to make the e.m.f. between the plates the same as that of the battery. In this state, the capacitor is said to be charged and an electric field will be present in the space between the two plates.

If, at some later time the switch is opened (see Fig. 2.24(c)), the positive plate will be left with a deficiency of electrons while the negative plate will be left with a surplus of electrons. Furthermore, since there is no path for current to flow between the two plates the capacitor will remain charged and a potential difference will be maintained between the plates. In practice, however, the stored charge will slowly decay due to the leakage resistance inside the capacitor.

Capacitance

The unit of capacitance is the farad (F). A capacitor is said to have a capacitance of 1 F if a current of 1 A flows in it when a voltage changing at the rate of 1 V/s is applied to it.

The current flowing in a capacitor will thus be proportional to the product of the capacitance (C) and the rate of change of applied voltage. Hence:

$i = C \times$ (rate of change of voltage)

The rate of change of voltage is often represented by the expression dv/dt where dv represents a very small change in voltage and dt represents the corresponding small change in time. Thus:

$$i = C \frac{dv}{dt}$$

Example 2.17

A voltage is changing at a uniform rate from 10 V to 50 V in a period of 0.1 s. If this voltage is applied to a capacitor of 22 μF, determine the current that will flow.

Solution

Now the current flowing will be given by:

$i = C \times$ (rate of change of voltage)

thus

$i = C\left(\dfrac{\text{change in voltage}}{\text{time}}\right) = 22 \times 10^{-6}\dfrac{50 - 10}{0.1}$

thus

$i = 22 \times 10^{-6} \times \dfrac{40}{0.1} = 22 \times 400 \times 10^{-6}$

$= 8.8 \times 10^{-3} = 8.8$ mA

Charge, capacitance and voltage

The charge or **quantity of electricity** that can be stored in the electric field between the capacitor plates is proportional to the applied voltage and the capacitance of the capacitor. Thus:

$Q = CV$

where Q is the charge (in coulombs), C is the capacitance (in farads), and V is the potential difference (in volts).

Example 2.18

A 10 µF capacitor is charged to a potential of 250 V. Determine the charge stored.

Solution

The charge stored will be given by:

$Q = CV = 10 \times 10^{-6} \times 250 = 2.5$ mC

Energy storage

The energy stored in a capacitor is proportional to the product of the capacitance and the square of the potential difference. Thus:

$E = 0.5CV^2$

where E is the energy (in joules), C is the capacitance (in farads), and V is the potential difference (in volts).

Example 2.19

A capacitor of 47 µF is required to store an energy of 4 J. Determine the potential difference which must be applied.

Solution

The foregoing formula can be re-arranged to make V the subject as follows:

$V = \left(\dfrac{E}{0.5C}\right)^{0.5} = \left(\dfrac{2E}{C}\right)^{0.5} = \left(\dfrac{2 \times 4}{47 \times 10^{-6}}\right)^{0.5}$

$= 130$ V

Capacitance and physical characteristics

The capacitance of a capacitor depends upon the physical dimensions of the capacitor (i.e. the size of the plates and the separation between them) and the dielectric material between the plates. The capacitance of a conventional parallel plate capacitor is given by:

$C = \dfrac{\varepsilon_o \varepsilon_r A}{d}$

where C is the capacitance (in farads), ε_o is the permittivity of free space, ε_r is the **relative permittivity** of the dielectric medium between the plates), and d is the separation between the plates (in metres).

Example 2.20

A capacitor of 1 nF is required. If a dielectric material of thickness 0.1 mm and relative permittivity 5.4 is available, determine the required plate area.

Solution

Re-arranging the formula $C = \varepsilon_o \varepsilon_r A/d$ to make A the subject gives:

$A = \dfrac{Cd}{\varepsilon_o \varepsilon_r} = \dfrac{1 \times 10^{-9} \times 0.1 \times 10^{-3}}{8.854 \times 10^{-12} \times 5.4}$

$= \dfrac{0.1 \times 10^{-12}}{47.8116 \times 10^{-12}}$

thus

$A = 0.002\ 09$ m^2 or 20.9 cm^2

In order to increase the capacitance of a capacitor, many practical components employ multiple plates (see Fig. 2.25). The capacitance is then given by:

$C = \dfrac{\varepsilon_o \varepsilon_r (n - 1)A}{d}$

where C is the capacitance (in farads), ε_o is the permittivity of free space, ε_r is the relative permittivity

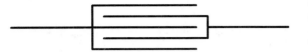

Figure 2.25   Multiple-plate capacitor

of the dielectric medium between the plates), d is the separation between the plates (in metres) and n is the total number of plates.

Example 2.21

A capacitor consists of six plates each of area 20 cm^2 separated by a dielectric of relative permittivity 4.5 and thickness 0.2 mm. Determine the capacitance of the capacitor.

Solution

Using $C = \varepsilon_o \varepsilon_r (n - 1) A/d$ gives:

$$C = \frac{8.854 \times 10^{-12}(6 - 1)\, 20 \times 10^{-4}}{0.2 \times 10^{-3}}$$

$$= \frac{885.40 \times 10^{-16}}{0.2 \times 10^{-3}}$$

$$C = 4427 \times 10^{-13} = 442.7 \times 10^{-12} \text{ F}$$

thus $C = 442.7$ pF.

Capacitor specifications

The specifications for a capacitor usually include the value of capacitance (expressed in microfarads, nanofarads or picofarads), the voltage rating (i.e. the maximum voltage which can be continuously applied to the capacitor under a given set of conditions), and the accuracy or tolerance (quoted as the maximum permissible percentage deviation from the marked value).

Other practical considerations when selecting capacitors for use in a particular application include temperature coefficient, leakage current, stability and ambient temperature range. Table 2.5 summarizes the properties of five of the most common types of capacitor. Figure 2.26 shows the construction of a typical tubular polystyrene capacitor.

Capacitor markings

The vast majority of capacitors employ written markings which indicate their values, working voltages, and tolerance. The most usual method of marking resin dipped polyester (and other) types of capacitor involves quoting the value (μF, nF or pF), the tolerance (often either 10% or 20%), and the working voltage (often using _ and ˜ to indicate d.c. and a.c., respectively). Several manufacturers

Table 2.5   Characteristics of common types of capacitor

Parameter	Capacitor type				
	Ceramic	Electrolytic	Metallized film	Mica	Polyester
Capacitance range (F)	2.2 p to 100 n	100 n to 68 m	1 μ to 16 μ	2.2 p to 10 n	10 n to 2.2 μ
Typical tolerance (%)	±10 and ±20	−10 to +50	±20	±1	±20
Typical voltage rating (d.c.)	50 V to 250 V	6.3 V to 400 V	250 V to 600 V	350 V	250 V
Temperature coefficient (ppm/°C)	+100 to −4700	+1000 typical	+100 to 200	+50	+250
Stability	Fair	Poor	Fair	Excellent	Good
Ambient temperature range (°C)	−85 to +85	−40 to +85	−25 to +85	−40 to +85	−40 to +100
Typical applications	Decoupling at high frequency	Smoothing and de-coupling at low frequency	Power supplies and power factor correction	Tuned circuits, filters, oscillators	General purpose

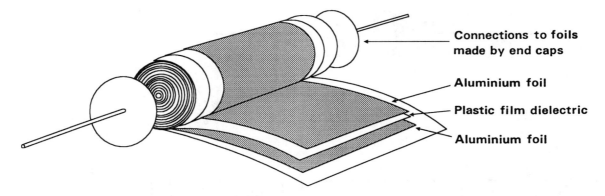

Figure 2.26   Construction of a typical tubular polystyrene capacitor

use two separate lines for their capacitor markings and these have the following meanings:

First line: capacitance (pF or μF) and tolerance (K = 10%, M = 20%)
Second line: rated d.c. voltage and code for the dielectric material

A three-digit code is commonly used to mark monolithic ceramic capacitors. The first two digits correspond to the first two digits of the value while the third digit is a multipler which gives the number of zeros to be added to give the value in picofarads.

Example 2.22

A monolithic ceramic capacitor is marked with the legend '103K'. What is its value?

Solution

The value (pF) will be given by the first two digits (10) followed by the number of zeros indicated by the third digit (3). The value of the capacitor is thus 10 000 pF or 10 nF. The final letter (K) indicates that the capacitor has a tolerance of 10%.

Some capacitors are marked with coloured stripes to indicate their value and tolerance (see Fig. 2.27).

Example 2.23

A tubular capacitor is marked with the following coloured stripes: brown, green, brown, red, brown. What is its value, tolerance, and working voltage?

Solution

See Fig. 2.28.

Series and parallel combination of capacitors

In order to obtain a particular value of capacitance, fixed capacitors may be arranged in either series or parallel (Figs 2.29 and 2.30).

The reciprocal of the effective capacitance of each of the series circuits shown in Fig. 2.29 is equal to the sum of the reciprocals of the individual capacitances. Hence, for Fig. 2.29(a)

$$\frac{1}{C} = \frac{1}{C_1} + \frac{1}{C_2}$$

while for Fig. 2.29(b)

$$\frac{1}{C} = \frac{1}{C_1} + \frac{1}{C_2} + \frac{1}{C_3}$$

In the former case, the formula can be more conveniently re-arranged as follows:

$$C = \frac{C_1 \times C_2}{C_1 + C_2}$$

(You can remember this as the *product* of the two capacitor values *divided by* the *sum* of the two values.)

For parallel arrangements of capacitors, the effective capacitance of the circuit is simply equal to the sum of the individual capacitances. Hence, for Fig. 2.30(a)

$$C = C_1 + C_2$$

while for Fig. 2.30(b)

$$C = C_1 + C_2 + C_3$$

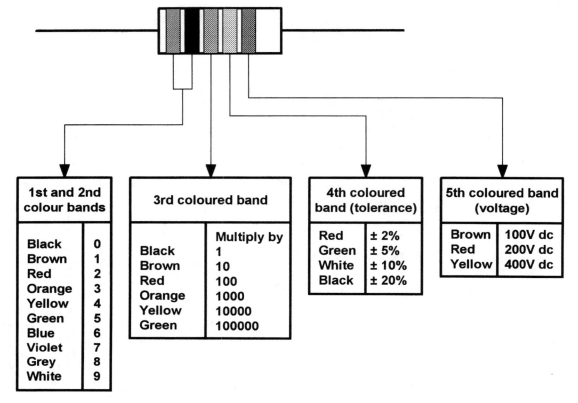

Figure 2.27   Capacitor colour code

Example 2.24

Determine the effective capacitance of the circuit shown in Fig. 2.31.

Solution

The circuit of Fig. 2.31 can be progressively simplified as shown in Fig. 2.32. The stages in this simplification are:

(a) C_1 and C_2 are in parallel and they can be replaced by a single capacitance (C_A) of (2 n + 4 n) = 6 n.
(b) C_A appears in parallel with C_3. These two capacitors can be replaced by a single capacitance (C_B) of (6 n × 2 n)/(6 n + 2 n) = 1.5 n.
(c) C_B appears in parallel with C_4. These two capacitors can be replaced by a single capacitance (C) of (1.5 n + 4 n) = 5.5 n.

Example 2.25

A capacitance of 50 nF (rated at 100 V) is required. What series combination of preferred value capacitors will satisfy this requirement? What voltage rating should each capacitor have?

Solution

Two 100 μF capacitors wired in series will provide a capacitance of 50 μF, as shown below:

$$C = \frac{C_1 \times C_2}{C_1 + C_2} = \frac{100\ \mu \times 100\ \mu}{100\ \mu + 100\ \mu} = \frac{10\ 000}{200} = 50\ \mu F$$

Since the capacitors are of equal value, the applied d.c. potential will be shared equally between them. Thus each capacitor should be rated at 50 V. Note that, in a practical circuit, we could take steps to ensure that the d.c. voltage was shared equally between the two capacitors by wiring equal, high-value (e.g. 100 kΩ) resistors across each capacitor.

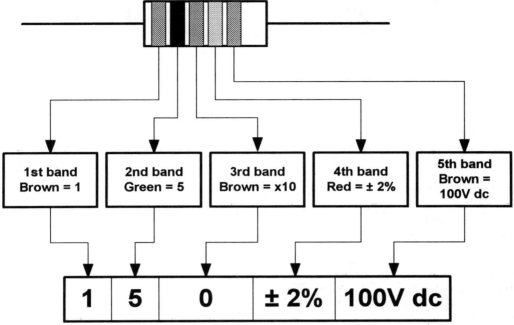

Brown, green, brown, red, brown

1st band Brown = 1	2nd band Green = 5	3rd band Brown = x10	4th band Red = ± 2%	5th band Brown = 100V dc

1	5	0	± 2%	100V dc

150pF ± 2% 100V dc

Figure 2.28

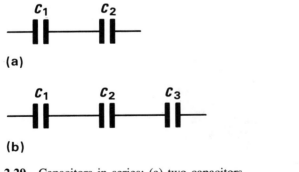

(a)

(b)

Figure 2.29 Capacitors in series: (a) two capacitors in series; (b) three capacitors in series

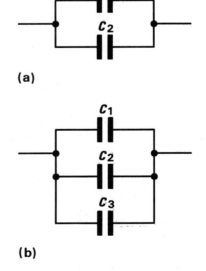

(a)

(b)

Figure 2.30 Capacitors in parallel: (a) two capacitors in parallel; (b) three capacitors in parallel

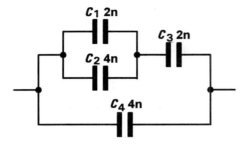

Figure 2.31   Circuit for Example 2.24

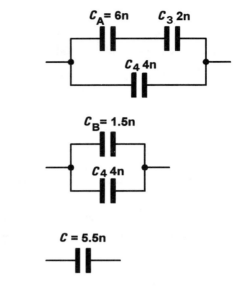

Figure 2.32   Stages in simplifying the circuit of Fig. 2.31

Variable capacitors

By moving one set of plates relative to the other, a capacitor can be made variable. The dielectric material used in a variable capacitor can be either air (see Fig. 2.33) or plastic (the latter tend to be more compact). Typical values for variable capacitors tend to range from about 25 pF to 500 pF. These components are commonly used for tuning radio receivers.

Inductors

Inductors provide us with a means of storing electrical energy in the form of a magnetic field. Typi-cal applications include chokes, filters and frequency selective circuits. The electrical characteristics of an inductor are determined by a number of factors including the material of the core (if any), the number of turns, and the physical dimensions of the coil. Figure 2.34 shows the construction of a basic air-cored inductor.

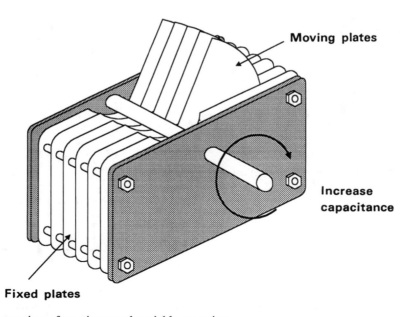

Figure 2.33   Construction of an air-spaced variable capacitor

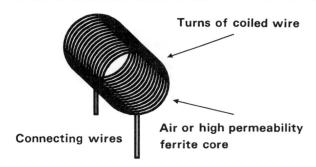

Figure 2.34 Basic air-cored inductor

Figure 2.35 A practical coil contains inductance **and** resistance

In practice every coil comprises both inductance (L) and resistance (R_S). The circuit of Fig. 2.35 shows these as two discrete components. In reality the inductance and resistance are both distributed throughout the component but it is convenient to treat the inductance and resistance as separate components in the analysis of the circuit.

Take a look at the circuit shown in Fig. 2.36(a). If the switch is left open, no current will flow and no magnetic flux will be produced by the inductor. If the switch is closed (see Fig. 2.36(b)), current will begin to flow as energy is taken from the supply in order to establish the magnetic field. However, the change in magnetic flux resulting from the appearance of current creates a voltage (an induced e.m.f.) across the coil which opposes the applied e.m.f. from the battery.

The induced e.m.f. results from the changing flux and it effectively prevents an instantaneous rise in current in the circuit. Instead, the current increases slowly to a maximum at a rate which depends upon the ratio of inductance (L) to resistance (R) present in the circuit. After a while, a steady state condition will be reached in which the voltage across the inductor will have decayed to zero and the current will have reached a maximum value (determined by the ratio of V to R, i.e. Ohm's law).

If, after this steady state condition has been achieved, the switch is opened (see Fig. 2.36(c)), the magnetic field will suddenly collapse and the energy will be returned to the circuit in the form of a 'back e.m.f.' which will appear across the coil as the field collapses.

Inductance

Inductance is the property of a coil which gives rise to the opposition to a change in the value of current flowing in it. Any change in the current applied to a coil/inductor will result in an induced voltage appearing across it.

The unit of inductance is the henry (H) and a coil is said to have an inductance of 1 H if a voltage of 1 V is induced across it when a current changing at the rate of 1 A/s is flowing in it.

The voltage induced across the terminals of an inductor will thus be proportional to the product of the inductance (L) and the rate of change of applied current. Hence:

$$e = -L \times \text{(rate of change of current)}$$

(Note that the minus sign indicates the polarity of the voltage, i.e. opposition to the change.)

Note that the rate of change of current is often represented by the expression di/dt where di represents a very small change in current and dt represents the corresponding small change in time.

$$e = -L\frac{di}{dt}$$

Example 2.26

A current increases at a uniform rate from 2 A to 6 A in a period of 250 ms. If this current is applied to an inductor of 600 mH, determine the voltage induced.

Solution

Now the induced voltage will be given by:

$$e = -L \times \text{(rate of change of current)}$$

thus

$$e = -L \times \left(\frac{\text{change in current}}{\text{time}}\right) = \frac{-0.6 \times (6-2)}{0.25}$$

$$= -9.6 \text{ V}$$

Energy storage

The energy stored in an inductor is proportional to the product of the inductance and the square of the current. Thus:

$$E = 0.5LI^2$$

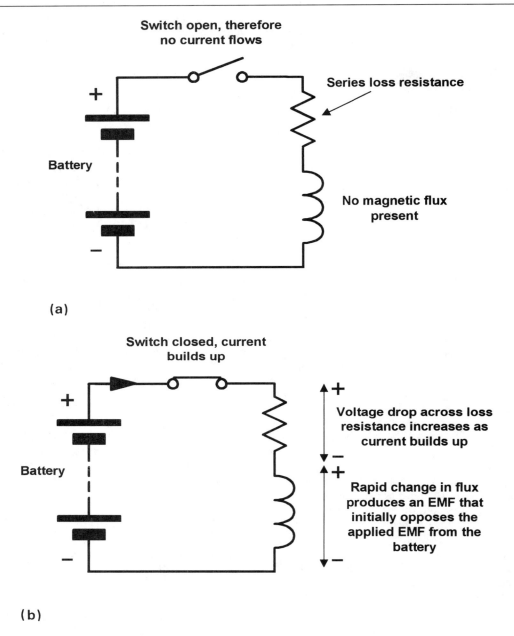

(a)

(b)

Figure 2.36 Effect of applying a current to an inductor: (a) initial state (no flux present); (b) flux rapidly builds up when current is applied; (c) flux collapses when current is removed

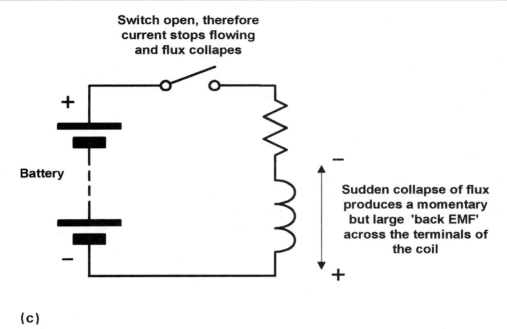

Switch open, therefore current stops flowing and flux collapes

+

Battery

−

Sudden collapse of flux produces a momentary but large 'back EMF' across the terminals of the coil

+

(c)

Figure 2.36   (Cont.)

where E is the energy (in joules), L is the inductance (in henries), and I is the current (in amperes).

Example 2.27

An inductor of 20 mH is required to store an energy of 2.5 J. Determine the current which must be applied.

Solution

The foregoing formula can be re-arranged to make I the subject as follows:

$$I = \left(\frac{E}{0.5L}\right)^{0.5} = \left(\frac{2E}{L}\right)^{0.5} = \left(\frac{2 \times 2.5}{20 \times 10^{-3}}\right)^{0.5} = \sqrt{250}$$

$$= 15.811 \text{ A}$$

Inductance and physical characteristics

The inductance of an inductor depends upon the physical dimensions of the inductor (e.g. the length and diameter of the winding), the number of turns, and the permeability of the material of the core. The inductance of an inductor is given by:

$$L = \frac{\mu_o\mu_r n^2 A}{l}$$

where L is the inductance (in henries), μ_o is the permeability of free space, μ_r is the relative permeability of the magnetic core, l is the length of the core (in metres), and A is the cross-sectional area of the core (in square metres).

Example 2.28

An inductor of 100 mH is required. If a closed magnetic core of length 20 cm, cross-sectional area 15 cm^2 and relative permeability 500 is available, determine the number of turns required.

Solution

Re-arranging the formula $L = \mu_o\mu_r n^2 A/l$ to make n the subject gives:

$$n = \frac{L \times l}{\mu_o\mu_r n^2 A} = \frac{0.1 \times 0.02}{12.57 \times 10^{-7} \times 20 \times 10^{-4}}$$

$$= \frac{0.002}{251.4 \times 10^{-11}}$$

thus

$$n = \frac{200}{251.4} \times 10^6 = 0.892 \times 10^3 = 892 \text{ turns}$$

Table 2.6 Characteristics of common types of inductor

Parameter	Inductor type			
	Air cored	Ferrite cored	Ferrite pot cored	Iron cored
Core material	Air	Ferrite	Ferrite	Iron
Inductance range (H)	50 n to 100 μ	10 μ to 1 m	1 m to 100 m	20 m to 20 H
Typical d.c. resistance (Ω)	0.05 to 10	1 to 100	2 to 100	10 to 200
Typical tolerance (%)	±10	±10	±10	±10
Typical Q-factor	60	80	40	20
Typical frequency range (Hz)	1 M to 500 M	100 k to 100 M	1 k to 10 M	50 Hz to 10 kHz
Typical applications	Tuned circuits	Filters and HF transformers	LF and MF filters and transformers	Smoothing chokes and LF filters

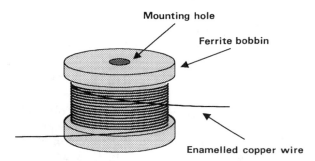

Figure 2.37 Construction of a typical ferrite cored inductor

Inductor specifications

Inductor specifications normally include the value of inductance (expressed in henries, millihenries or microhenries), the current rating (i.e. the maximum current which can be continuously applied to the inductor under a given set of conditions), and the accuracy or tolerance (quoted as the maximum permissible percentage deviation from the marked value). Other considerations may include the temperature coefficient of the inductance (usually expressed in parts per million, ppm, per unit temperature change), the stability of the inductor, the d.c. resistance of the coil windings (ideally zero), the Q-factor (quality factor) of the coil, and the recommended working frequency range. Table 2.6 summarizes the properties of four common types of inductor. Figure 2.37 shows the construction of a typical ferrite-cored inductor.

Inductor markings

As with capacitors, the vast majority of inductors use written markings to indicate values, working current, and tolerance. Some small inductors are marked with coloured stripes to indicate their value and tolerance (the standard colour values are used and inductance is normally expressed in microhenries).

Series and parallel combinations of inductors

In order to obtain a particular value of inductance, fixed inductors may be arranged in either series or parallel as shown in Figs 2.38 and 2.39. The effective inductance of each of the series circuits shown in Fig. 2.38 is simply equal to the sum of the individual inductances. Hence, for Fig. 2.38(a)

$$L = L_1 + L_2$$

while for Fig. 2.38(b)

$$L = L_1 + L_2 + L_3$$

Turning to the parallel inductors shown in Fig. 2.39, the reciprocal of the effective inductance of each circuit is equal to the sum of the reciprocals of the individual inductances. Hence, for Fig. 2.39(a)

$$\frac{1}{L} = \frac{1}{L_1} + \frac{1}{L_2}$$

while for Fig. 2.39(b)

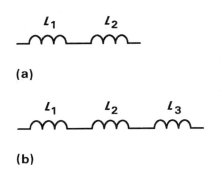

(a)

(b)

Figure 2.38 Inductors in series: (a) two inductors in series (b) three inductors in series

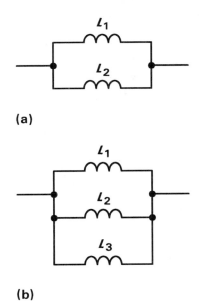

(a)

(b)

Figure 2.39 Inductors in parallel: (a) two inductors in parallel (b) three inductors in parallel

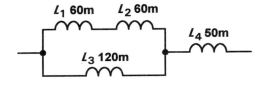

Figure 2.40 Circuit for Example 2.29

$$\frac{1}{L} = \frac{1}{L_1} + \frac{1}{L_2} + \frac{1}{L_3}$$

In the former case, the formula can be more conveniently re-arranged as follows:

$$L = \frac{L_1 \times L_2}{L_1 + L_2}$$

(You can remember this as the *product* of the two inductance values *divided by* the *sum* of the two values.)

Example 2.29

An inductance of 5 mH (rated at 2 A) is required. What parallel combination of preferred value inductors will satisfy this requirement?

Solution

Two 10 mH inductors may be wired in parallel to provide an inductance of 5 mH as shown below:

$$L = \frac{L_1 \times L_2}{L_1 + L_2} = \frac{10\,\text{m} \times 10\,\text{m}}{10\,\text{m} + 10\,\text{m}} = \frac{100\,\text{m}}{20\,\text{m}} = 5\,\text{mH}$$

Since the inductors are identical, the applied current will be shared equally between them. Hence each inductor should have a current rating of 1 A.

Example 2.30

Determine the effective inductance of the circuit shown in Fig. 2.40.

Solution

The circuit can be progressively simplified as shown in Fig. 2.41. The stages in this simplification are:

(a) L_1 and L_2 are in series and they can be replaced by a single inductance (L_A) of (60 m + 60 m) = 120 mH.
(b) L_A appears in parallel with L_3. These two inductors can be replaced by a single inductance (L_B) of (120 m × 120 m)/(120 m + 120 m) = 60 mH.
(c) L_B appears in series with L_4. These two inductors can be replaced by a single inductance (L) of (60 m + 50 m) = 110 mH.

Variable inductors

A ferrite cored inductor can be made variable by moving its core in or out of the former onto which

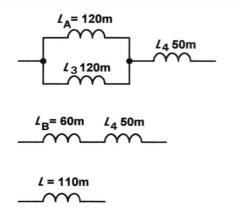

Figure 2.41   Stages in simplifying the circuit of Fig. 2.40

the coil is wound. Many small inductors have threaded ferrite cores to make this possible. Such inductors are often used in radio and high-frequency applications where precise tuning is required.

Formulae introduced in this chapter

Component tolerance:
(page 20)

$$\text{Tolerance} = \frac{\text{error}}{\text{marked value}} \times 100\%$$

Resistors in series:
(page 23)

$$R = R_1 + R_2 + R_3$$

Resistors in parallel:
(page 23)

$$\frac{1}{R} = \frac{1}{R_1} + \frac{1}{R_2} + \frac{1}{R_3}$$

Two resistors in parallel:
(page 23)

$$R = \frac{R_1 \times R_2}{R_1 + R_2}$$

Resistance and temperature:
(page 25)

$$R_t = R_0(1 + \alpha t)$$

Current flowing in a capacitor:
(page 32)

$$i = C \frac{dv}{dt}$$

Charge stored in a capacitor:
(page 33)

$$Q = CV$$

Energy stored in a capacitor:
(page 33)

$$E = \frac{1}{2} CV^2$$

Capacitance of a capacitor:
(page 33)

$$C = \frac{\varepsilon_0 \varepsilon_r A}{d}$$

Capacitors in series:
(page 35)

$$\frac{1}{C} = \frac{1}{C_1} + \frac{1}{C_2} + \frac{1}{C_3}$$

Two capacitors in series:
(page 35)

$$C = \frac{C_1 \times C_2}{C_1 + C_2}$$

Capacitors in parallel:
(page 35)

$$C = C_1 + C_2 + C_3$$

Induced e.m.f. in an inductor:
(page 39)

$$e = -L \frac{di}{dt}$$

Energy stored in an inductor:
(page 39)

$$E = \frac{1}{2}LI^2$$

Inductance of an inductor:
(page 41)

$$L = \frac{\mu_0 \mu_r n^2 A}{l}$$

Inductors in series:
(page 42)

$$L = L_1 + L_2 + L_3$$

Circuit symbols introduced in this chapter

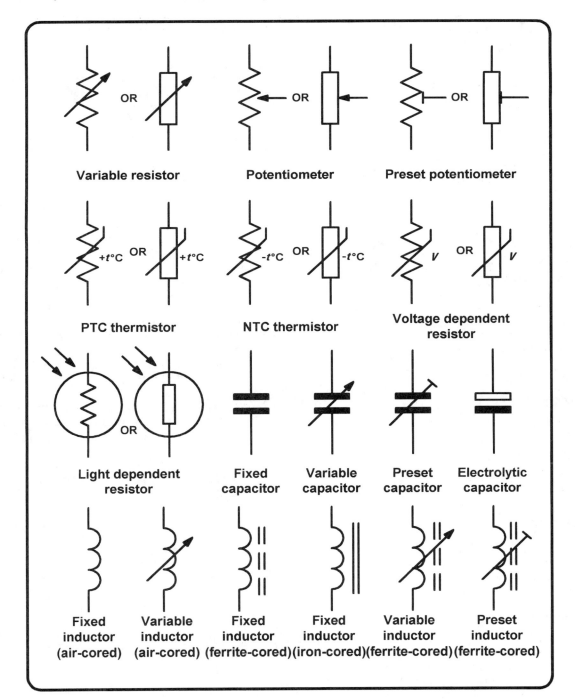

Figure 2.42

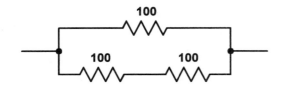

Figure 2.43

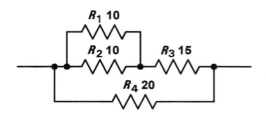

Figure 2.44

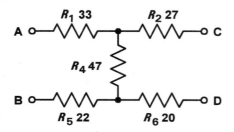

Figure 2.45

Inductors in parallel:
(page 43)

$$\frac{1}{L} = \frac{1}{L_1} + \frac{1}{L_2} + \frac{1}{L_3}$$

Two inductors in parallel:
(page 43)

$$L = \frac{L_1 \times L_2}{L_1 + L_2}$$

Problems

2.1   A power supply rated at 15 V, 0.25 A is to be tested at full rated output. What value of load resistance is required and what power rating should it have? What type of resistor is most suitable for this application and why?

2.2   Determine the value and tolerance of resistors marked with the following coloured bands:

(a) red, violet, yellow, gold;
(b) brown, black, black, silver;
(c) blue, grey, green, gold;
(d) orange, white, silver, gold;
(e) red, red, black, brown, red.

2.3   A batch of resistors are all marked yellow, violet, black, gold. If a resistor is selected from this batch within what range would you expect its value to be?

2.4   Resistors of 27 Ω, 33 Ω, 56 Ω and 68 Ω are available. How can two or more of these resistors be arranged to realize the following resistance values:

(a) 60 Ω
(b) 14.9 Ω
(c) 124 Ω
(d) 11.7 Ω
(e) 128 Ω.

2.5   Three 100 Ω resistors are connected as shown in Fig. 2.43. Determine the effective resistance of the circuit.

2.6   Determine the effective resistance of the circuit shown in Fig. 2.44.

2.7   Determine the resistance of the network shown in Fig. 2.45 looking into terminals A and B with (a) terminals C and D open-circuit and (b) terminals C and D short-circuit.

2.8   A resistor has a temperature coefficient of 0.0008/°C. If the resistor has a resistance of 390 Ω at 0°C, determine its resistance at 55°C.

2.9   A resistor has a temperature coefficient of 0.004/°C. If the resistor has a resistance of 82 kΩ at 20°C, what will its resistance be at 75°C?

2.10   A resistor has a resistance of 218 Ω at 0°C and 225 Ω at 100°C. Determine the resistor's temperature coefficient.

2.11   Capacitors of 1 μF, 3.3 μF, 4.7 μF and 10 μF are available. How can two or more of these capacitors be arranged to realize the following capacitance values:

(a) 8 μF
(b) 11 μF
(c) 19 μF
(d) 0.91 μF
(e) 1.94 μF.

2.12   Three 180 pF capacitors are connected (a) in series and (b) in parallel. Determine the effective capacitance in each case.

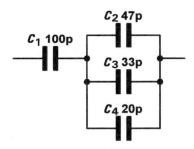

Figure 2.46

2.13 Determine the effective capacitance of the circuit shown in Fig. 2.46.

2.14 A capacitor of 330 μF is charged to a potential of 63 V. Determine the quantity of charge stored.

2.15 A parallel plate capacitor has plates of area 0.02 m^2. Determine the capacitance of the capacitor if the plates are separated by a dielectric of thickness 0.5 mm and relative permittivity 5.6.

2.16 A capacitor is required to store 0.5 J of energy when charged from a 120 V d.c. supply. Determine the value of capacitor required.

2.17 The current in a 2.5 H inductor increases uniformly from zero to 50 mA in 400 ms. Determine the induced e.m.f.

2.18 An inductor comprises 200 turns of wire wound on a closed magnetic core of length 24 cm, cross-sectional area 10 cm^2 and relative permeability 650. Determine the inductance of the inductor.

2.19 A current of 4 A flows in a 60 mH inductor. Determine the energy stored.

2.20 Inductors of 10 mH, 22 mH, 60 mH and 100 mH are available. How can two or more of these inductors be arranged to realize the following inductance values:

(a) 6.2 mH
(b) 6.9 mH
(c) 32 mH
(d) 70 mH
(e) 170 mH.

(Answers to these problems appear on page 202.)

3

D.C. circuits

In many cases, Ohm's law alone is insufficient to determine the magnitude of the voltages and currents present in a circuit. This chapter introduces several techniques that simplify the task of solving complex circuits. It also introduces the concept of exponential growth and decay of voltage and current in circuits containing capacitance and resistance and inductance and resistance. It concludes by showing how humble C–R and L–R circuits can be used for shaping the waveforms found in electronic circuits. We start by introducing two of the most useful laws of electronics.

Kirchhoff's laws

Kirchhoff's laws relate to the algebraic sum of currents at a junction (or node) or voltages in a network (or mesh). The term 'algebraic' simply indicates that the polarity of each current or voltage drop must be taken into account by giving it an appropriate sign, either positive (+) or negative (−).

Kirchhoff's current law states that the algebraic sum of the currents present at a junction (node) in a circuit is zero (see Fig. 3.1).

Kirchhoff's voltage law states that the algebraic sum of the potential drops in a closed network (or 'mesh') is zero (see Fig. 3.2).

Example 3.1

Determine the currents and voltages in the circuit of Fig. 3.3.

Solution

In order to solve the circuit shown in Fig. 3.3, it is first necessary to identify the currents and voltages as shown in Figs 3.4 and 3.5.

By applying Kirchhoff's current law at node A in Fig. 3.4:

$$I_1 + I_2 = I_3 \tag{1}$$

By applying Kirchhoff's voltage law in loop A of Fig. 3.5:

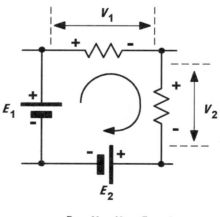

$$E_1 - V_1 - V_2 - E_2 = 0$$

Convention:
Move clockwise around the circuit starting with the positive terminal of the largest EMF.
Voltages acting in the same sense are positive (+)
Voltages acting in the opposite sense are negative (-)

Figure 3.2 Kirchhoff's voltage law

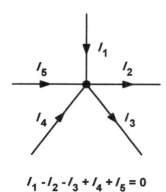

$$I_1 - I_2 - I_3 + I_4 + I_5 = 0$$

Convention:
Currents flowing towards the junction are positive (+)
Currents flowing away from the junction are negative (-)

Figure 3.1 Kirchhoff's current law

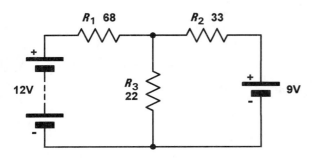

Figure 3.3

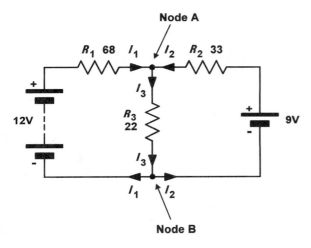

Figure 3.4   Currents in Example 3.1

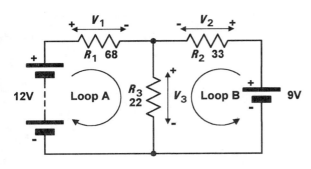

Figure 3.5   Voltages in Example 3.1

$$12 = V_1 + V_3 \qquad (2)$$

By applying Kirchhoff's voltage law in loop B of Fig. 3.5:

$$9 = V_2 + V_3 \qquad (3)$$

By applying Ohm's law:

$$V_1 = I_1 R_1 = I_1 \times 68 \qquad (4)$$

$$V_2 = I_2 R_2 = I_2 \times 33 \qquad (5)$$

and

$$V_3 = I_3 R_3 = I_3 \times 22 \qquad (6)$$

From (1) and (6):

$$V_3 = (I_1 + I_2) \times 22 \qquad (7)$$

Combining (4), (7) and (2):

$$12 = I_1 \times 68 + (I_1 + I_2) \times 22$$

or

$$12 = 68I_1 + 22I_1 + 22I_2$$

thus

$$12 = 90I_1 + 22I_2 \qquad (8)$$

Combining (4), (7) and (3):

$$9 = I_2 \times 33 + (I_1 + I_2) \times 22 \qquad (9)$$

or

$$9 = 33I_2 + 22I_1 + 22I_2$$

thus

$$9 = 22I_1 + 55I_2 \qquad (10)$$

Multiplying (8) by 5 gives:

$$60 = 450I_1 + 110I_2 \qquad (11)$$

Multiplying (10) by 2 gives:

$$18 = 44I_1 + 110I_2 \qquad (12)$$

Subtracting (12) from (11):

$$60 - 18 = 450I_1 - 44I_1$$

$$42 = 406I_1$$

thus

$$I_1 = 42/406 = 0.103 \text{ A}$$

From (8):

$$12 = 90 \times 0.103 + 22I_2$$

thus

$$12 = 9.27 + 22I_2$$

or

$$2.73 = 22I_2$$

thus

$$I_2 = 2.73/22 = 0.124 \text{ A}$$

From (1):

$$I_3 = I_1 + I_2$$

thus

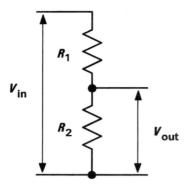

Figure 3.6   Potential divider circuit

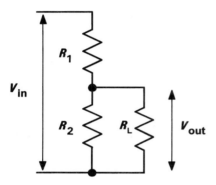

Figure 3.7   Loaded potential divider

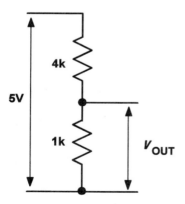

Figure 3.8

$I_3 = 0.103 + 0.124 = 0.227$ A

From (4):

$V_1 = 0.103 \times 68 = 7$ V

From (5):

$V_2 = 0.124 \times 33 = 4$ V

From (6):

$V_3 = 0.227 \times 22 = 5$ V

Check from (2):

$12 = V_1 + V_3 = 7 + 5 = 12$

Check from (3):

$9 = V_2 + V_3 = 4 + 5 = 9$

The potential divider

The potential divider circuit (see Fig. 3.6) is commonly used to reduce voltage levels in a circuit. The output voltage produced by the circuit is given by:

$$V_{out} = V_{in} \times \frac{R_2}{R_1 + R_2}$$

It is, however, important to note that the output voltage (V_{out}) will fall when current is drawn from the arrangement. Figure 3.7 shows the effect of **loading** the potential divider circuit.

In the loaded potential divider (Fig. 3.7) the output voltage is given by:

$$V_{out} = V_{in} \times \frac{R_p}{R_1 + R_p}$$

where

$$R_p = \frac{(R_2 \times R_L)}{(R_2 + R_L)}$$

Example 3.2

The potential divider shown in Fig. 3.8 is used as a simple voltage calibrator. Determine the output voltage produced by the circuit:

(a) when the output terminals are left open-circuit (i.e. no load is connected); and
(b) when the output is loaded by a resistance of 10 kΩ.

Solution

(a) In the first case we can simply apply the formula:

$$V_{out} = V_{in} \times \frac{R_2}{R_1 + R_2}$$

where $V_{in} = 5$ V, $R_1 = 4$ kΩ and $R_2 = 1$ kΩ. Hence

$$V_{out} = 5 \times \frac{1}{4 + 1} = 5 \times \frac{1}{5} = 1 \text{ V}$$

(b) In the second case we need to take into account the effect of the 10 kΩ resistor connected to the output terminals of the potential divider.

First we need to find the equivalent resistance of the parallel combination of R_2 and R_L:

$$R_p = \frac{(R_2 \times R_L)}{(R_2 + R_L)} = \frac{(1 \text{ k}\Omega \times 10 \text{ k}\Omega)}{(1 \text{ k}\Omega + 10 \text{ k}\Omega)}$$
$$= 10/11 = 0.909 \text{ k}\Omega$$

Then we can determine the output voltage from:

$$V_{out} = V_{in} \times \frac{R_p}{R_1 + R_p} = 5 \times \frac{0.909}{4 + 0.909}$$
$$= 5 \times \frac{0.909}{4.909}$$

Hence

$$V_{out} = 5 \times 0.185 = 0.925 \text{ V}$$

The current divider

The current divider circuit (see Fig. 3.9) is used to divert current from one branch of a circuit to another. The output current produced by the circuit is given by:

$$I_{out} = I_{in} \times \frac{R_1}{R_1 + R_2}$$

It is, however, important to note that the output current (I_{out}) will fall when the load connected to the output terminals has any appreciable resistance.

Example 3.3

A moving coil meter requires a current of 1 mA to provide full-scale deflection. If the meter coil has a resistance of 100 Ω and is to be used as a milliammeter reading 5 mA full-scale, determine the value of parallel shunt resistor required.

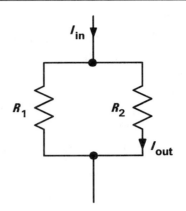

Figure 3.9   Current divider circuit

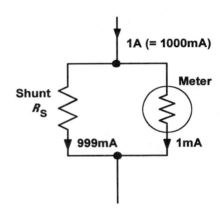

Figure 3.10

Solution

This problem may sound a little complicated so it is worth taking a look at the equivalent circuit of the meter (Fig. 3.10) and comparing it with the current divider shown in Fig. 3.9.

We can apply the current divider formula, replacing I_{out} with I_m (the meter full-scale deflection current) and R_2 with R_m (the meter resistance). R_1 is the required value of shunt resistor, R_s. Hence:

$$I_m = I_{in} \times \frac{R_s}{R_s + R_m}$$

Re-arranging the formula gives:

$$I_m \times (R_s + R_m) = I_{in} \times R_s$$

thus

$$I_m R_s + I_m R_m = I_{in} R_s$$

or

$$I_{in} R_s - I_m R_s = I_m R_m$$

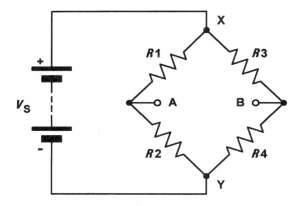

Figure 3.11   Basic Wheatstone bridge arrangement

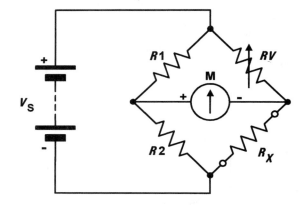

Figure 3.12   Practical Wheatstone bridge

hence

$R_s(I_{in} - I_m) = I_m R_m$

and

$$R_s = \frac{I_m R_m}{(I_{in} - I_m)}$$

Now $I_m = 1$ mA, $R_m = 100 \Omega$, $I_{in} = 5$ mA, thus

$$R_s = \frac{1 \text{ mA} \times 100 \text{ }\Omega}{(5 \text{ mA} - 1 \text{ mA})} = \frac{1}{4} \times 100 \text{ }\Omega = 25 \text{ }\Omega$$

The Wheatstone bridge

The Wheatstone bridge forms the basis of a number of electronic circuits including several that are used in instrumentation and measurement. The basic form of Wheatstone bridge is shown in Fig. 3.11. The voltage developed between A and B will be zero when the voltage between A and Y is the same as that between B and Y. In effect, $R1$ and $R2$ constitute a potential divider as do $R3$ and $R4$. The bridge will be **balanced** (and $V_{AB} = 0$) when the ratio of $R1:R2$ is the same as the ratio $R3:R4$. Hence, at balance:

$R1/R2 = R3/R4$

A practical form of Wheatstone bridge that can be used for measuring unknown resistances is shown in Fig. 3.12. $R1$ and $R2$ constitute the two **ratio arms** while one arm (that occupied by $R3$ in Fig. 3.11) is replaced by a calibrated variable resistor. The unknown resistor, R_x, is connected in the fourth arm.

At balance:

$R1/R2 = RV/R_x$ thus $R_x = R2/R1 \times RV$

Example 3.4

A Wheatstone bridge is based on the circuit shown in Fig. 3.12. If $R1$ and $R2$ can each be switched so that they have values of either 100Ω or $1 \text{ k}\Omega$ and RV is variable between 10Ω and $10 \text{ k}\Omega$, determine the range of resistance values that can be measured.

Solution

The maximum value of resistance that can be measured will correspond to the largest ratio of $R2:R1$ (i.e. when $R2$ is $1 \text{ k}\Omega$ and $R1$ is 100Ω) and the highest value of RV (i.e. $10 \text{ k}\Omega$). In this case:

$R_x = R2/R1 \times RV = 1 \text{ k}\Omega/100 \text{ }\Omega \times 10 \text{ k}\Omega$
$= 10 \times 10 \text{ k}\Omega = 100 \text{ k}\Omega$

The minimum value of resistance that can be measured will correspond to the smallest ratio of $R2:R1$ (i.e. when $R2$ is 100Ω and $R1$ is $1 \text{ k}\Omega$) and the smallest value of RV (i.e. 10Ω). In this case:

$R_x = R2/R1 \times RV = 100 \text{ }\Omega/1 \text{ k}\Omega \times 10 \text{ }\Omega$
$= 0.1 \times 10 \text{ }\Omega = 1 \text{ }\Omega$

Hence the range of values that can be measured extends from 1Ω to $100 \text{ k}\Omega$.

Thevenin's theorem

Thevenin's theorem allows us to replace a complicated network of resistances and voltage sources with a simple equivalent circuit comprising a single **voltage source** connected in series with a single resistance (see Fig. 3.13).

The single voltage source in the Thevenin equivalent circuit, V_{OC}, is simply the voltage that appears

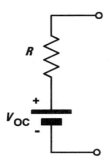

Figure 3.13 Thevenin equivalent circuit

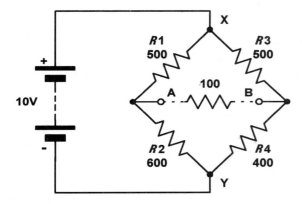

Figure 3.14

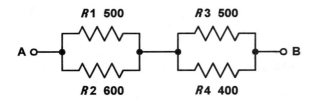

Figure 3.15 Determining the equivalent resistance in Fig. 3.14

between the terminals when nothing is connected to it. In other words, it is the *open-circuit* voltage that would appear between X and Y.

The single resistance that appears in the Thevenin equivalent circuit, R, is the resistance that would be seen *looking into* the network between X and Y when all of the voltage sources (assumed perfect) are replaced by *short-circuit* connections. Note that if the voltage sources are not perfect (i.e. if they have some internal resistance) the equivalent circuit must be constructed on the basis that each voltage source is replaced by its own internal resistance.

Once we have values for V_{OC} and R, we can determine how the network will behave when it is connected to a load (i.e. when a resistor is connected across the terminals X and Y).

Example 3.5

Figure 3.14 shows a Wheatstone bridge. Determine the current that will flow in a 100 Ω load connected between terminals A and B.

Solution

First we need to find the Thevenin equivalent of the circuit. To find V_{OC} we can treat the bridge arrangement as two potential dividers. The voltage across $R2$ and $R4$ will be given by:
 For $R2$

$$V = 10 \times \frac{R2}{R1 + R2} = 10 \times \frac{600}{500 + 600}$$
$$= 10 \times 0.5454 = 5.454 \text{ V}$$

Hence the voltage at A, relative to Y, will be 5.454 V.
 For $R4$

$$V = 10 \times \frac{R4}{R3 + R4} = 10 \times \frac{400}{500 + 400}$$
$$= 10 \times 0.4444 = 4.444 \text{ V}$$

Hence the voltage at B, relative to Y, will be 4.444 V.

The voltage V_{AB} will be the difference between V_{AY} and V_{BY}, hence the open-circuit output voltage, V_{AB}, will be given by:

$$V_{AB} = V_{AY} - V_{BY} = 5.454 - 4.444 = 1.01 \text{ V}$$

Next we need to find the Thevenin equivalent resistance. To do this, we can redraw the circuit, replacing the battery with a short-circuit, as shown in Fig. 3.15.

The equivalent resistance is given by:

$$R = \frac{R1 \times R2}{R1 + R2} + \frac{R3 \times R4}{R3 + R4}$$
$$= \frac{500 \times 600}{500 + 600} + \frac{500 \times 400}{500 + 400}$$

thus

$$R = \frac{300\,000}{1100} + \frac{200\,000}{900} = 272.7 + 222.2$$
$$= 494.9 \ \Omega$$

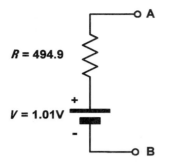

Figure 3.16   Thevenin equivalent of the circuit in Fig. 3.14

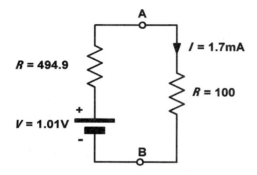

Figure 3.17   Determining the current when the Thevenin equivalent circuit is loaded

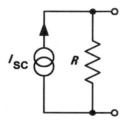

Figure 3.18   Norton equivalent circuit

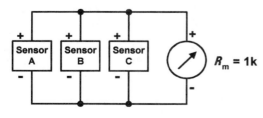

Figure 3.19

The Thevenin equivalent circuit is shown in Fig. 3.16. To determine the current in a 100 Ω load connected between A and B, we can make use of the Thevenin equivalent circuit by simply adding a 100 Ω resistor to the circuit and applying Ohm's law, as shown in Fig. 3.17.

The current flowing in Fig. 3.17 will be given by:

$$I = \frac{V}{R + 100} + \frac{10}{494.9 + 100} = \frac{10}{594.9} = 0.0168 \text{ A}$$

$$= 16.8 \text{ mA}$$

Norton's theorem

Norton's theorem provides an alternative method of reducing a complex network to a simple equivalent circuit. Unlike Thevenin's theorem, Norton's theorem makes use of a **current source** rather than a voltage source. The Norton equivalent circuit allows us to replace a complicated network of resistances and voltage sources with a simple equivalent circuit comprising a single constant current source connected in parallel with a single resistance (see Fig. 3.18).

The constant current source in the Norton equivalent circuit, I_{SC}, is simply the *short-circuit* current that would flow if X and Y were to be linked directly together.

The resistance that appears in the Norton equivalent circuit, R, is the resistance that would be seen *looking into* the network between X and Y when all of the voltage sources are replaced by *short-circuit* connections. Once again, it is worth noting that, if the voltage sources have any appreciable internal resistance, the equivalent circuit must be constructed on the basis that each voltage source is replaced by its own internal resistance.

As with the Thevenin equivalent, we can determine how a network will behave by obtaining values for I_{SC} and R.

Example 3.6

Three parallel connected temperature sensors having the following characteristics are connected in parallel as shown in Fig. 3.19.

Sensor	A	B	C
Output voltage (open circuit)	20 mV	30 mV	10 mV
Internal resistance	5 kΩ	3 kΩ	2 kΩ

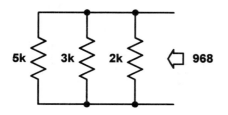

Figure 3.20 Determining the equivalent resistance in Fig. 3.19

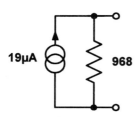

Figure 3.21 Norton equivalent of the circuit in Fig. 3.19

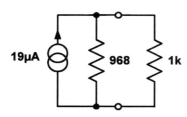

Figure 3.22 Determining the voltage when the Norton equivalent circuit is loaded

Determine the voltage produced when the arrangement is connected to a moving-coil meter having a resistance of 1 kΩ.

Solution

First we need to find the Norton equivalent of the circuit. To find I_{SC} we can determine the short-circuit current from each sensor and add them together.

For sensor A

$$I = \frac{V}{R} + \frac{20 \text{ mV}}{5 \text{ k}\Omega} = 4 \text{ }\mu\text{A}$$

For sensor B

$$I = \frac{V}{R} + \frac{30 \text{ mV}}{3 \text{ k}\Omega} = 10 \text{ }\mu\text{A}$$

For sensor C

$$I = \frac{V}{R} + \frac{10 \text{ mV}}{2 \text{ k}\Omega} = 5 \text{ }\mu\text{A}$$

The total current, I_{SC}, will be given by:

$$I_{SC} = 4 \text{ }\mu\text{A} + 10 \text{ }\mu\text{A} + 5 \text{ }\mu\text{A} = 19 \text{ }\mu\text{A}$$

Next we need to find the Norton equivalent resistance. To do this, we can redraw the circuit showing each sensor replaced by its internal resistance, as shown in Fig. 3.20.

The equivalent resistance is given by:

$$\frac{1}{R} = \frac{1}{R1} + \frac{1}{R2} + \frac{1}{R3} = \frac{1}{5 \text{ k}\Omega} + \frac{1}{3 \text{ k}\Omega} + \frac{1}{2 \text{ k}\Omega}$$
$$= 0.0002 + 0.000 \text{ } 33 + 0.0005$$

thus

$$\frac{1}{R} = 0.001 \text{ } 03 \text{ or } R = 968 \text{ }\Omega$$

The Norton equivalent circuit is shown in Fig. 3.21. To determine the voltage in a 1 kΩ moving-coil meter connected between A and B, we can make use of the Norton equivalent circuit by simply adding a 1 kΩ resistor to the circuit and applying Ohm's law, as shown in Fig. 3.22.

The voltage appearing across the moving-coil meter in Fig. 3.22 will be given by:

$$V = I_{sc} \times \frac{R \times R_M}{R + R_M} = 19 \text{ }\mu\text{A} \times \frac{1000 \times 968}{1000 + 968}$$
$$= 19 \text{ }\mu\text{A} \times 492 \text{ }\Omega$$

or $V = 9.35$ mV.

C–R circuits

Networks of capacitors and resistors (known as C–R circuits) form the basis of many timing and pulse shaping circuits and are thus often found in practical electronic circuits.

Charging

A simple C–R circuit is shown in Fig. 3.23. When the network is connected to a constant voltage source (V_s), as shown in Fig. 3.24, the voltage (v_c) across the (initially uncharged) capacitor voltage will rise exponentially as shown in Fig. 3.25. At the same time, the current in the circuit (i) will fall, as shown in Fig. 3.26.

The rate of growth of voltage with time (and

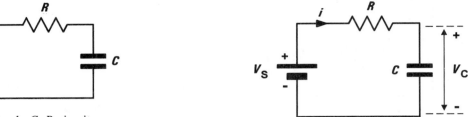

Figure 3.23   Simple *C–R* circuit

Figure 3.24   *C–R* circuit (*C* charging through *R*)

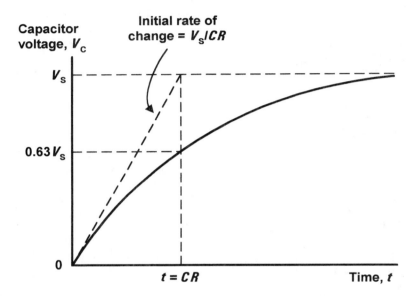

Figure 3.25   Exponential growth of capacitor voltage (v_C) in Fig. 3.24

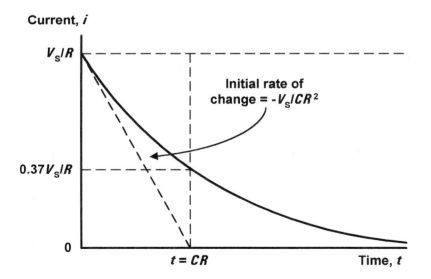

Figure 3.26   Exponential decay of current (*i*) in Fig. 3.24

decay of current with time) will be dependent upon the product of capacitance and resistance. This value is known as the **time constant** of the circuit. Hence:

time constant, $t = C \times R$

where C is the value of capacitance (F), R is the resistance (Ω), and t is the time constant (s).

The voltage developed across the charging capacitor (v_c) varies with time (t) according to the relationship:

$$v_c = V_s(1 - e^{-t/CR})$$

where v_c is the capacitor voltage, V_s is the d.c. supply voltage, t is the time, and CR is the time constant of the circuit (equal to the product of capacitance, C, and resistance, R).

The capacitor voltage will rise to approximately 63% of the supply voltage in a time interval equal to the time constant. At the end of the next interval of time equal to the time constant (i.e. after an elapsed time equal to $2CR$) the voltage will have risen by 63% of the remainder, and so on.

In theory, the capacitor will **never** become fully charged. However, after a period of time equal to $5CR$, the capacitor voltage will to all intents and purposes be equal to the supply voltage. At this point the capacitor voltage will have risen to 99.3% of its final value and we can consider it to be fully charged.

During charging, the current in the capacitor (i) varies with time (t) according to the relationship:

$$i = V_s\, e^{-t/CR}$$

where v_c is the capacitor voltage, V_s is the supply voltage, t is the time, C is the capacitance, and R is the resistance.

The current will fall to approximately 37% of the initial current in a time equal to the time constant. At the end of the next interval of time equal to the time constant (i.e. after a total time of $2CR$ has elapsed) the current will have fallen by a further 37% of the remainder, and so on.

Example 3.7

An initially uncharged capacitor of 1 µF is charged from a 9 V d.c. supply via a 3.3 MΩ resistor. Determine the capacitor voltage 1 s after connecting the supply.

Solution

The formula for exponential growth of voltage in the capacitor is:

$$v_c = V_s(1 - e^{-t/CR})$$

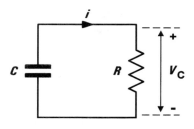

Figure 3.27 *C–R circuit (C discharges through R)*

where $V_s = 9$ V, $t = 1$ s and $CR = 1\ \mu\text{F} \times 3.3\ \text{M}\Omega = 3.3$ s. Thus

$$v_c = 9(1 - e^{-1/3.3})$$

or

$$v_c = 9(1 - 0.738)$$

hence

$$v_c = 9 \times 0.262 = 2.358 \text{ V}$$

Discharge

Having considered the situation when a capacitor is being charged, let's consider what happens when an already charged capacitor is discharged. When the fully charged capacitor from Fig. 3.24 is connected as shown in Fig. 3.27, the capacitor will discharge through the resistor, and the capacitor voltage (v_c) will fall exponentially with time, as shown in Fig. 3.28. The current in the circuit (i) will also fall, as shown in Fig. 3.29. The rate of discharge (i.e. the rate of decay of voltage with time) will once again be governed by the time constant of the circuit ($C \times R$).

The voltage developed across the discharging capacitor (v_c) varies with time (t) according to the relationship:

$$v_c = V_s\, e^{-t/CR}$$

The capacitor voltage will fall to approximately 37% of the initial voltage in a time equal to the time constant. At the end of the next interval of time equal to the time constant (i.e. after an elapsed time equal to $2CR$) the voltage will have fallen by 37% of the remainder, and so on.

In theory, the capacitor will **never** become fully discharged. After a period of time equal to $5CR$, however, the capacitor voltage will to all intents and purposes be zero. At this point the capacitor voltage will have fallen below 1% of its initial value. At this point we can consider it to be fully discharged.

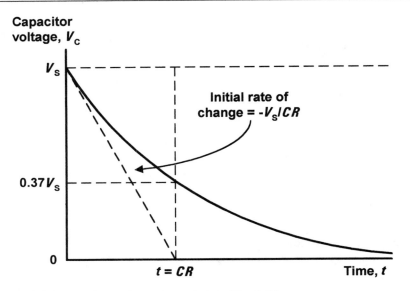

Figure 3.28   Exponential decay of capacitor voltage (v_C) in Fig. 3.27

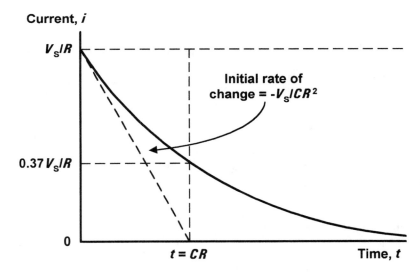

Figure 3.29   Exponential decay of current (i) in Fig. 3.27

As with charging, the current in the capacitor (i) varies with time (t) according to the relationship:

$$i = V_s\, e^{-t/CR}$$

where v_c is the capacitor voltage, V_s is the supply voltage, t is the time, C is the capacitance, and R is the resistance.

The current will fall to approximately 37% of the initial voltage in a time equal to the time constant. At the end of the next interval of time equal to the time constant (i.e. after a total time of $2CR$ has elapsed) the voltage will have fallen by a further 37% of the remainder, and so on.

Example 3.8

A 10 μF capacitor is charged to a potential of 20 V and then discharged through a 47 kΩ resistor. Determine the time taken for the capacitor voltage to fall below 10 V.

Table 3.1   Exponential growth and decay

$t/(CR)$ or $t/(L/R)$	K^a growth	decay
0.0	0.0000	1.0000
0.1	0.0951	0.9048
0.2	0.1812	0.8187 (1)
0.3	0.2591	0.7408
0.4	0.3296	0.6703
0.5	0.3935	0.6065
0.6	0.4511	0.5488
0.7	0.5034	0.4965
0.8	0.5506	0.4493
0.9	0.5934	0.4065
1.0	0.6321	0.3679
1.5	0.7769	0.2231
2.0	0.8647 (2)	0.1353
2.5	0.9179	0.0821
3.0	0.9502	0.0498
3.5	0.9698	0.0302
4.0	0.9817	0.0183
4.5	0.9889	0.0111
5.0	0.9933	0.0067

Notes: (1) See Example 3.9
       (2) See Example 3.13
[a] K is the ratio of instantaneous value to final value (i.e. v_c/V_s etc.).

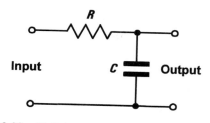

Figure 3.30 C–R integrating circuit

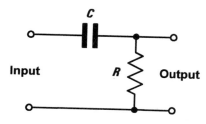

Figure 3.31 C–R differentiating circuit

Solution

The formula for exponential decay of voltage in the capacitor is:

$$v_c = V_s\, e^{-t/CR}$$

where $V_s = 20$ V and $CR = 10\ \mu\text{F} \times 47\ \text{k}\Omega = 0.47$ s.
  We need to find t when $v_c = 10$ V.
  Rearranging the formula to make t the subject gives:

$$t = -CR \times \ln(v_c/V_s)$$

thus

$$t = -0.47 \times \ln(10/20)$$

or

$$t = -470 \times -0.693 = 0.325\ \text{s}$$

In order to simplify the mathematics of exponential growth and decay, Table 3.1 provides an alternative tabular method that may be used to determine the voltage and current in a C–R circuit.

Example 3.9

A 150 μF capacitor is charged to a potential of 150 V. The capacitor is then removed from the charging source and connected to a 2 MΩ resistor. Determine the capacitor voltage 1 minute later.

Solution

We will solve this problem using Table 3.1 rather than the exponential formula.
  First we need to find the time constant:

$$C \times R = 150\ \mu\text{F} \times 2\ \text{M}\Omega = 300\ \text{s}$$

Next we find the ratio of t to CR:
  After 1 minute, $t = 60$ s therefore the ratio of t to CR is 60/300 or 0.2. Table 3.1 shows that when $t/CR = 0.2$, the ratio of instantaneous value to final value (K) is 0.8187.
  Thus

$$v_c/V_s = 0.8187$$

or

$$v_c = 0.8187 \times V_s = 0.8187 \times 150\ \text{V} = 122.8\ \text{V}$$

Waveshaping with C–R networks

One of the most common applications of C–R networks is in waveshaping circuits. The circuits shown in Figs 3.30 and 3.31 function as simple square-to-

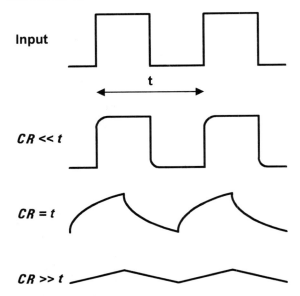

Figure 3.32 Waveforms for the integrating circuit (Fig. 3.30)

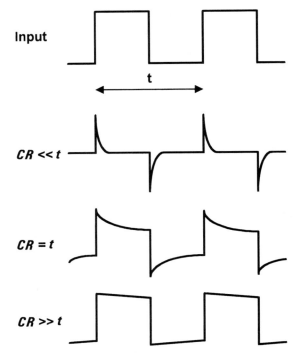

Figure 3.33 Waveforms for the differentiating circuit (Fig. 3.31)

triangle and square-to-pulse converters by, respectively, **integrating** and **differentiating** their inputs.

The effectiveness of the simple **integrator circuit** shown in Fig. 3.30 depends very much upon the ratio of time constant $(C \times R)$ to periodic time (t). The larger this ratio is, the more effective the circuit will be as an integrator. The effectiveness of the circuit of Fig. 3.30 is illustrated by the input and output waveforms shown in Fig. 3.32.

Similarly, the effectiveness of the simple **differentiator circuit** shown in Fig. 3.31 also depends very much upon the ratio of time constant $(C \times R)$ to periodic time (t). The smaller this ratio is, the more effective the circuit will be as a differentiator. The effectiveness of the circuit of Fig. 3.31 is illustrated by the input and output waveforms shown in Fig. 3.33.

Example 3.10

A circuit is required to produce a train of alternating positive and negative pulses of short duration from a square wave of frequency 1 kHz. Devise a suitable C–R circuit and specify suitable values.

Solution

Here we require the services of a differentiating circuit along the lines of that shown in Fig. 3.31. In order that the circuit operates effectively as a differentiator, we need to make the time constant

$(C \times R)$ very much less than the periodic time of the input waveform (1 ms). Assuming that we choose a medium value for R of, say, 10 kΩ, the maximum value which we could allow C to have would be that which satisfies the equation:

$$C \times R = 0.1t$$

where $R = 10$ kΩ and $t = 1$ ms. Thus

$$C = \frac{0.1t}{R} = \frac{0.1 \times 1 \text{ ms}}{10 \text{ k}\Omega} = 0.1 \times 10^{-3} \times 10^{-4}$$

$$= 1 \times 10^{-8}$$

or

$$C = 10 \times 10^{-9} = 10 \text{ nF}$$

In practice, any value equal to or less than 10 nF would be adequate. A very small value (say below 1 nF) will, however, generate pulses of a very narrow width.

Example 3.11

A circuit is required to produce a triangular waveform from a square wave of frequency 1 kHz. Devise a suitable C–R arrangement and specify suitable values.

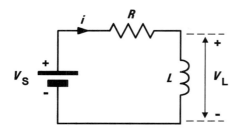

Figure 3.34   *L–R* circuit

Solution

This time we require an integrating circuit like that shown in Fig. 3.32. In order that the circuit operates effectively as an integrator, we need to make the time constant ($C \times R$) very much greater than the period time of the input waveform (1 ms). Assuming that we choose a medium value for R of, say, 10 kΩ, the minimum value which we could allow C to have would be that which satisfies the equation:

$$C \times R = 10t$$

where $R = 10$ kΩ and $t = 1$ ms. Thus

$$C = \frac{10t}{R} = \frac{10 \times 1 \text{ ms}}{10 \text{ k}\Omega} = 10 \times 10^{-3} \times 10^{-4}$$
$$= 10 \times 10^{-7}$$

or

$$C = 1 \times 10^{-6} = 1 \text{ } \mu\text{F}$$

In practice, any value equal to or greater than 1 μF would be adequate. A very large value (say above 10 μF) will, however, produce a triangle wave of very severely limited amplitude.

L–R circuits

Networks of inductors and resistors (known as *L–R* circuits) can also be used for timing and pulse shaping. In comparison with capacitors, however, inductors are somewhat more difficult to manufacture and are consequently more expensive. Inductors are also prone to losses and may also require screening to minimize the effects of stray magnetic coupling. Inductors are, therefore, generally unsuited to simple timing and waveshaping applications.

Figure 3.34 shows a simple *LR* network in which an inductor is connected to a constant voltage supply. When the supply is first connected, the current (i) will rise exponentially with time (as shown in Fig. 3.35). At the same time, the inductor voltage (V_L) will fall (as shown in Fig. 3.36). The rate of change of current with time will depend upon the ratio of inductance to resistance and is known as the time constant. Hence:

time constant, $t = L/R$

where L is the value of inductance (H), R is the resistance (Ω), and t is the time constant (s).

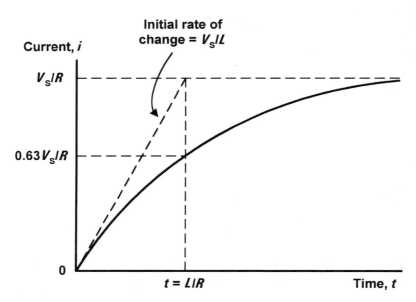

Figure 3.35   Exponential growth of current (i) in Fig. 3.34

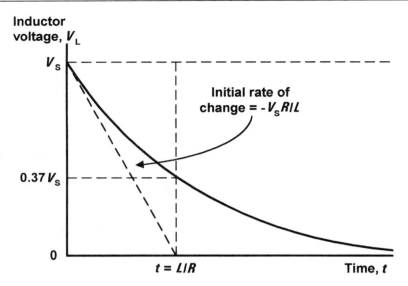

Figure 3.36   Exponential decay of inductor voltage (v_L) in Fig. 3.34

The current flowing in the inductor (i) varies with time (t) according to the relationship:

$i = V_s/R(1 - e^{-tR/L})$

where V_s is the d.c. supply voltage, R is the resistance of the inductor, and L is the inductance.

The current (i) will initially be zero and will rise to approximately 63% of its maximum value (i.e. V_s/R) in a time interval equal to the time constant. At the end of the next interval of time equal to the time constant (i.e. after a total time of $2L/R$ has elapsed) the current will have risen by a further 63% of the remainder, and so on.

In theory, the current in the inductor will *never* become equal to V_s/R. However, after a period of time equal to $5L/R$, the current will to all intents and purposes be equal to V_s/L. At this point the current in the inductor will have risen to 99.3% of its final value.

The voltage developed across the inductor (V_L) varies with time (t) according to the relationship:

$V_L = V_s\, e^{-tR/L}$

where V_s is the d.c. supply voltage, R is the resistance of the inductor, and L is the inductance.

The inductor voltage will fall to approximately 37% of the initial voltage in a time equal to the time constant. At the end of the next interval of time equal to the time constant (i.e. after a total time of $2L/R$ has elapsed) the voltage will have fallen by a further 37% of the remainder, and so on.

Example 3.12

A coil having inductance 6 H and resistance 24 Ω is connected to a 12 V d.c. supply. Determine the current in the inductor 0.1 s after the supply is first connected.

Solution

The formula for exponential growth of current in the coil is:

$i = V_s/R(1 - e^{-tR/L})$

where $V_s = 12$ V and $L/R = 6$ H$/24$ $\Omega = 0.25$ s.
We need to find i when $t = 0.1$ s

$i = 12/24(1 - e^{-0.1/0.25}) = 0.5(1 - e^{-0.4}) = 0.5(1 - 0.67)$

thus

$i = 0.5 \times 0.33 = 0.165$ A.

In order to simplify the mathematics of exponential growth and decay, Table 3.1 provides an alternative tabular method that may be used to determine the voltage and current in an L–R circuit.

Example 3.13

A coil has an inductance of 100 mH and a resistance of 10 Ω. If the inductor is connected to a 5 V

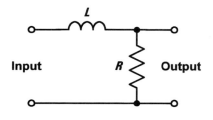

Figure 3.37   *L–R* integrating circuit

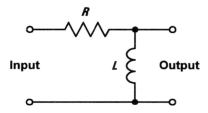

Figure 3.38   *L–R* differentiating circuit

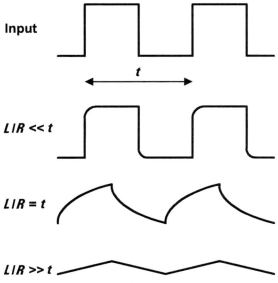

Figure 3.39   Waveforms for the integrating circuit (Fig. 3.37)

d.c. supply, determine the inductor voltage 20 ms after the supply is first connected.

Solution

We will solve this problem using Table 3.1 rather than the exponential formula.

First we need to find the time constant:

$$L/R = 0.1 \text{ H}/10 \text{ } \Omega = 0.01 \text{ s.}$$

Next we find the ratio of t to L/R:

when $t = 20$ ms the ratio of t to L/R is 0.02/0.01 or 2.

Table 3.1 shows that when $t/(L/R) = 2$, the ratio of instantaneous value to final value (K) is 0.8647. Thus

$$v_L/V_s = 0.8647$$

or

$$v_L = 0.8647 \times v_s = 0.8647 \times 5 \text{ V} = 4.32 \text{ V}$$

Waveshaping with *L–R* networks

L–R networks are sometimes employed in waveshaping applications. The circuits shown in Figs 3.37 and 3.38 function as square-to-triangle and square-to-pulse converters by, respectively, integrating and differentiating their inputs.

The effectiveness of the simple **integrator circuit** shown in Fig. 3.37 depends very much upon the ratio of time constant (L/R) to periodic time (t). The larger this ratio is, the more effective the circuit will be as an integrator. The effectiveness of the circuit of Fig. 3.37 is illustrated by the input and output waveforms shown in Fig. 3.39.

Similarly, the effectiveness of the simple **differentiator** circuit shown in Fig. 3.38 also depends very much upon the ratio of time constant (L/R) to periodic time (t). The smaller this ratio is, the more effective the circuit will be as a differentiator. The effectiveness of the circuit of Fig. 3.38 is illustrated by the input and output waveforms shown in Fig. 3.40.

In practical waveshaping applications, *C–R* circuits are almost invariably superior to *L–R* circuits on the grounds of both cost and performance. Hence examples of the use of *LR* circuits in waveshaping applications have not been given.

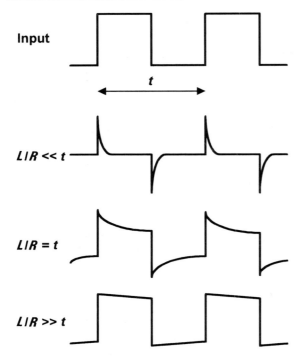

Figure 3.40   Waveforms for the differentiating circuit (Fig. 3.38)

Formulae introduced in this chapter

Kirchhoff's current law:
(page 48)

Algebraic sum of currents = 0

Kirchhoff's voltage law:
(page 48)

Algebraic sum of e.m.f.s = algebraic sum of voltage drops

Potential divider:
(page 50)

$$V_{out} = V_{in} \times \frac{R_2}{R_1 + R_2}$$

Current divider:
(page 51)

$$I_{out} = I_{in} \times \frac{R_1}{R_1 + R_2}$$

Wheatstone bridge:
(page 52)

$R1/R2 = R3/R4$
$R_x = RV \times R2/R1$

Time constant of a *C–R* circuit:
(page 57)

$t = CR$

Capacitor voltage (charge):
(page 57)

$v_c = V_s(1 - e^{-t/CR})$

Capacitor current (charge):
(page 57)

$i = V_s\, e^{-t/CR}$

Capacitor voltage (discharge):
(page 57)

$v_c = V_s\, e^{-t/CR}$

Capacitor current (discharge):
(page 58)

$i = V_s\, e^{-t/CR}$

Time constant of an *L–R* circuit:
(page 61)

$t = L/R$

Inductor current (flux build up):
(page 62)

$i = V_s/R(1 - e^{-tR/L})$

Inductor voltage (flux build up):
(page 62)

$V_L = V_s\, e^{-tR/L}$

Problems

3.1   A power supply is rated at 500 mA maximum output current. If the supply delivers 150 mA to one circuit and 75 mA to another, how much current would be available for a third circuit?

3.2   A 15 V d.c. supply delivers a total current of 300 mA. If this current is shared equally between four circuits, determine the resistance of each circuit.

3.3   Determine the unknown current in each circuit shown in Fig. 3.42.

3.4   Determine the unknown voltage in each circuit shown in Fig. 3.43.

3.5   Determine all currents and voltages in Fig. 3.44.

3.6   Two resistors, one of 120 Ω and one of 680 Ω, are connected as a potential divider across a 12 V supply. Determine the voltage developed across each resistor.

3.7   Two resistors, one of 15 Ω and one of 5 Ω, are connected in parallel. If a current of 2 A

Circuit symbols introduced in this chapter

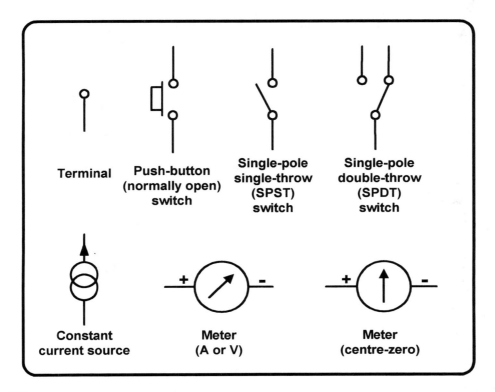

Figure 3.41

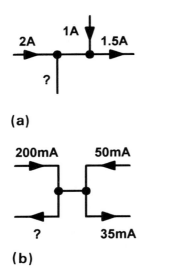

(a)

(b)

Figure 3.42

is applied to the combination, determine the current flowing in each resistor.

3.8   A switched attenuator comprises five 1 kΩ resistors wired in series across a 5 V d.c. supply. If the output voltage is selected by means of a single-pole four-way switch, determine the voltage produced for each switch position.

3.9   A battery charger is designed to charge six 9 V batteries simultaneously from a raw 24 V d.c. supply. Each battery is connected to the 24 V supply via a separate resistor. If the batteries are to be charged at a nominal current of 10 mA each, determine the value of series resistance and the total drain on the 24 V supply.

3.10   A capacitor of 1 μF is charged from a 15 V d.c. supply via a 100 kΩ resistor. How long will it take for the capacitor voltage to reach 5 V?

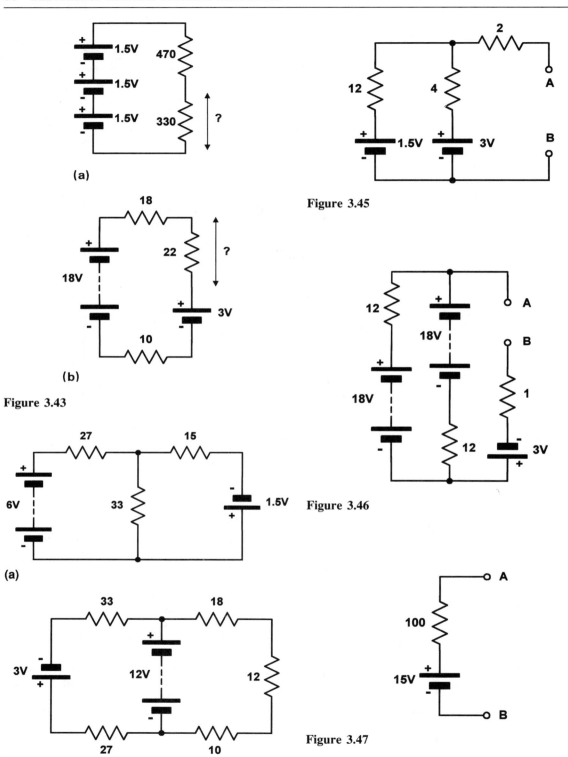

(a)

Figure 3.45

(b)

Figure 3.43

Figure 3.46

(a)

(b)

Figure 3.44

Figure 3.47

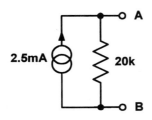

Figure 3.48

3.11   A capacitor of 22 µF is charged to a voltage of 50 V. If the capacitor is then discharged using a resistor of 100 kΩ, determine the time taken for the capacitor voltage to reach 10 V.

3.12   An initially uncharged capacitor is charged from a 200 V d.c. supply through a 2 MΩ resistor. If it takes 50 s for the capacitor voltage to reach 100 V, determine the value of capacitance.

3.13   An inductor has an inductance of 2.5 H and a resistance of 10 Ω. If the inductor is connected to a 5 V d.c. supply, determine the time taken for the current to grow to 200 mA.

3.14   Determine the Thevenin equivalent of the circuit shown in Fig. 3.45.

3.15   Determine the Norton equivalent of the circuit shown in Fig. 3.46.

3.16   The Thevenin equivalent of a network is shown in Fig. 3.47. Determine (a) the short-circuit output current and (b) the output voltage developed across a load of 200 Ω.

3.17   The Norton equivalent of a network is shown in Fig. 3.48. Determine (a) the open-circuit output voltage and (b) the output voltage developed across a load of 5 kΩ.

(Answers to these problems appear on page 202.)

4

Alternating voltage and current

This chapter introduces basic alternating current theory. We discuss the terminology used to describe alternating waveforms and the behaviour of resistors, capacitors, and inductors when an alternating current is applied to them. The chapter concludes by introducing another useful component, the transformer.

AC versus d.c.

Direct currents are currents which, even though their magnitude may vary, essentially flow only in one direction. In other words, direct currents are unidirectional. Alternating currents, on the other hand, are bidirectional and continuously reverse their direction of flow. The polarity of the e.m.f. which produces an alternating current must consequently

also be changing from positive to negative, and vice versa.

Alternating currents produce alternating potential differences (voltages) in the circuits in which they flow. Furthermore, in some circuits, alternating voltages may be superimposed on direct voltage levels (see Fig. 4.1). The resulting voltage may be unipolar (i.e. always positive or always negative) or bipolar (i.e. partly positive and partly negative).

Waveforms and signals

A graph showing the variation of voltage or current present in a circuit is known as a **waveform**. There are many common types of waveform encountered in electrical circuits including sine (or sinusoidal), square, triangle, ramp or sawtooth

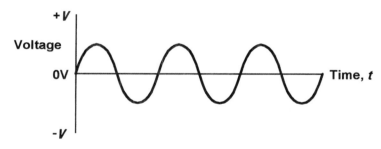

(a)

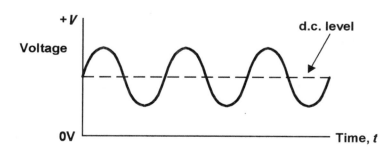

(b)

Figure 4.1 (a) Bipolar sine wave (this waveform swings symmetrically above and below 0 V); (b) unipolar sine wave (this waveform is superimposed on a d.c. level)

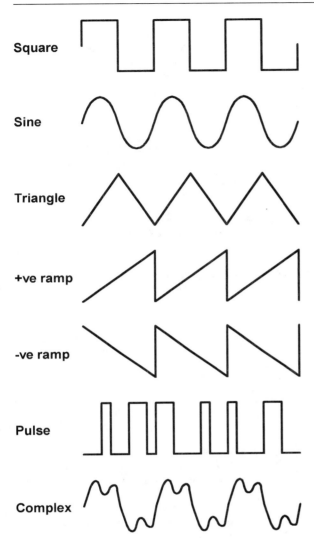

Square	
Sine	
Triangle	
+ve ramp	
-ve ramp	
Pulse	
Complex	

Figure 4.2   Common waveforms

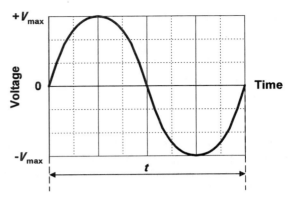

Figure 4.3   One cycle of a sine wave voltage showing its **periodic time**

either **analogue** (continuously variable) or **digital** (based on discrete states).

Frequency

The frequency of a repetitive waveform is the number of cycles of the waveform which occur in unit time. Frequency is expressed in hertz (Hz). A frequency of 1 Hz is equivalent to one cycle per second. Hence, if a voltage has a frequency of 400 Hz, 400 cycles will occur in every second.

The equation for the voltage shown in Fig. 4.3 at a time, t, is:

$$v = V_{max} \sin(2\pi ft)$$

where v is the instantaneous voltage, V_{max} is the maximum (or peak) voltage and f is the frequency.

Example 4.1

A sine wave voltage has a maximum value of 20 V and a frequency of 50 Hz. Determine the instantaneous voltage present (a) 2.5 ms and (b) 15 ms from the start of the cycle.

Solution

We can determine the voltage at any instant of time using:

$$v = V_{max} \sin(2\pi ft)$$

where $V_{max} = 20$ V and $f = 50$ Hz.

In (a), $t = 2.5$ ms, hence:

$$v = 20 \sin(2\pi \times 50 \times 0.0025) = 20 \sin (0.785)$$
$$= 20 \times 0.707 = 14.14 \text{ V}$$

(which may be either positive or negative going), and pulse. **Complex waveforms** like speech or music usually comprise many components at different frequencies. **Pulse waveforms** are often categorized as either repetitive or non-repetitive (the former comprises a pattern of pulses which regularly repeats while the latter comprises pulses which constitute a unique event). Several of the most common waveform types are shown in Fig. 4.2.

Signals can be conveyed using one or more of the properties of a waveform and sent using wires, cables, optical and radio links. Signals can also be processed in various ways using amplifiers, modulators, filters, etc. Signals are also classified as

In (b), $t = 15$ ms, hence:

$v = 20 \sin(2\pi \times 50 \times 0.015) = 20 \sin (4.71)$
$= 20 \times -1 = -20$ V

Periodic time

The periodic time (or period) of a waveform is the time take for one complete cycle of the wave (see Fig. 4.3). The relationship between periodic time and frequency is thus:

$t = 1/f$ or $f = 1/t$

where t is the periodic time (in s) and f is the frequency (in Hz).

Example 4.2

A waveform has a frequency of 400 Hz. What is the periodic time of the waveform?

Solution

$t = 1/f = 1/400 = 0.0025$ s (or 2.5 ms)

Hence the waveform has a periodic time of 2.5 ms.

Example 4.3

A waveform has a periodic time of 40 ms. What is its frequency?

Solution

$$f = 1/t = \frac{1}{40 \times 10^{-3}} = \frac{1}{0.04} = 25 \text{ Hz}$$

Average, peak, peak–peak, and r.m.s. values

The **average value** of an alternating current which swings symmetrically above and below zero will obviously be zero when measured over a long period of time. Hence average values of currents and voltages are invariably taken over one complete half-cycle (either positive or negative) rather than over one complete full-cycle (which would result in an average value of zero).

The **amplitude** (or **peak value**) of a waveform is a measure of the extent of its voltage or current

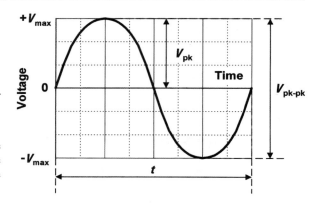

Figure 4.4   One cycle of a sine wave voltage showing its **peak** and **peak–peak** values

Table 4.1   Multiplying factors for average, peak, peak–peak and r.m.s. values

Given quantity	Wanted quantity			
	Average	Peak	Peak–peak	r.m.s.
Average	1	1.57	3.14	1.11
Peak	0.636	1	2	0.707
Peak–peak	0.318	0.5	1	0.353
r.m.s.	0.9	1.414	2.828	1

excursion from the resting value (usually zero). The **peak-to-peak value** for a wave which is symmetrical about its resting value is twice its peak value (see Fig. 4.4).

The **r.m.s.** (or **effective**) **value** of an alternating voltage or current is the value which would produce the same heat energy in a resistor as a direct voltage or current of the same magnitude. Since the r.m.s. value of a waveform is very much dependent upon its shape, values are only meaningful when dealing with a waveform of known shape. Where the shape of a waveform is not specified, r.m.s. values are normally assumed to refer to sinusoidal conditions.

For a given waveform, a set of fixed relationships exist between average, peak, peak–peak, and r.m.s. values. The required multiplying factors are summarized for sinusoidal voltages and currents in Table 4.1.

Example 4.4

A sinusoidal voltage has an r.m.s. value of 240 V. What is the peak value of the voltage?

Solution

The corresponding multiplying factor (found from Table 4.1) is 1.414. Hence:

$V_{pk} = 1.414 \times V_{r.m.s.} = 1.414 \times 240 = 339.4$ V

Example 4.5

An alternating current has a peak–peak value of 50 mA. What is its r.m.s. value?

Solution

The corresponding multiplying factor (found from Table 4.1) is 0.353. Hence:

$I_{r.m.s.} = 0.353 \times I_{pk-pk} = 0.353 \times 0.05 = 0.0177$ A
      (or 17.7 mA)

Example 4.6

A sinusoidal voltage 10 V pk–pk is applied to a resistor of 1 kΩ. What value of r.m.s. current will flow in the resistor?

Solution

This problem must be solved in two stages. First we will determine the peak–peak current in the resistor and then we shall convert this value into a corresponding r.m.s. quantity. Since

$I = V/R$, $I_{pk-pk} = V_{pk-pk}/R$

Hence:

$I_{pk-pk} = 10$ $V_{pk-pk}/1$ kΩ $= 10$ mA$_{pk-pk}$

The required multiplying factor (peak–peak to r.m.s.) is 0.353. Thus:

$I_{r.m.s.} = 0.353 \times I_{pk-pk} = 0.353 \times 10$ mA $= 3.53$ mA

Reactance

When alternating voltages are applied to capacitors or inductors the magnitude of the current flowing will depend upon the value of capacitance or inductance and on the frequency of the voltage. In effect, capacitors and inductors oppose the flow of current in much the same way as a resistor. The important difference being that the effective resistance (or reactance) of the component varies with frequency (unlike the case of a conventional resis-

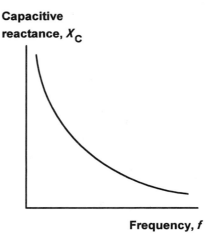

Figure 4.5   Variation of reactance with frequency for a capacitor

tor where the magnitude of the current does not change with frequency).

Capacitive reactance

The reactance of a capacitor is defined as the ratio of applied voltage to current and, like resistance, it is measured in ohms. The reactance of a capacitor is inversely proportional to both the value of capacitance and the frequency of the applied voltage. Capacitive reactance can be found by applying the following formula:

$$X_C = \frac{V_C}{I_C} = \frac{1}{2\pi f C}$$

where X_C is the reactance in ohms, f is the frequency in hertz, and C is the capacitance in Farads.

Capacitive reactance falls as frequency increases, as shown in Fig. 4.5.

The applied voltage, V_C, and current, I_C, flowing in a pure capacitive reactance will differ in phase by an angle of 90° or π/2 radians (the **current leads the voltage**). This relationship is illustrated in the current and voltage waveforms (drawn to a common time scale) shown in Fig. 4.6 and as a phasor diagram shown in Fig. 4.7.

Example 4.7

Determine the reactance of a 1 μF capacitor at (a) 100 Hz and (b) 10 kHz.

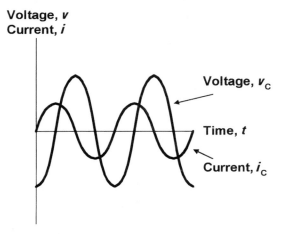

Figure 4.6   Voltage and current waveforms for a pure capacitor (the current leads the voltage by 90°)

Solution

(a) At 100 Hz

$$X_C = \frac{1}{2 \times \pi \times 100 \times 1 \times 10^{-6}}$$

or

$$X_C = \frac{0.159}{10^{-4}} = 0.159 \times 10^4$$

Thus

$$X_C = 1.59 \text{ k}\Omega$$

(b) At 10 kHz

$$X_C = \frac{1}{2 \times \pi \times 10\,000 \times 1 \times 10^{-6}}$$

or

$$X_C = \frac{0.159}{10^{-2}} = 0.159 \times 10^2$$

Thus

$$X_C = 15.9 \ \Omega$$

Example 4.8

A 100 nF capacitor is to form part of a filter connected across a 240 V 50 Hz mains supply. What current will flow in the capacitor?

Solution

First we must find the reactance of the capacitor:

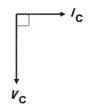

Figure 4.7   Phasor diagram for a pure capacitor

$$X_C = \frac{1}{2 \times \pi \times 50 \times 100 \times 10^{-9}} = 31.8 \text{ k}\Omega$$

The r.m.s. current flowing in the capacitor will thus be:

$$I_C = \frac{V_C}{X_C} = \frac{240 \text{ V}}{31.8 \text{ k}\Omega} = 7.5 \text{ mA}$$

Inductive reactance

The reactance of an inductor is defined as the ratio of applied voltage to current and, like resistance, it is measured in ohms. The reactance of an inductor is directly proportional to both the value of inductance and the frequency of the applied voltage. Inductive reactance can be found by applying the formula:

$$X_L = \frac{V_L}{I_L} = 2\pi f L$$

where X_L is the reactance in Ω, f is the frequency in Hz, and L is the inductance in H.

Inductive reactance increases linearly with frequency as shown in Fig. 4.8.

The applied voltage current, I_L, and voltage, V_L, developed across a pure inductive reactance will differ in phase by an angle of 90° or $\pi/2$ radians (the **current lags the voltage**). This relationship is illustrated in the current and voltage waveforms (drawn to a common time scale) shown in Fig. 4.9 and as a phasor diagram shown in Fig. 4.10.

Example 4.9

Determine the reactance of a 10 mH inductor at (a) 100 Hz and (b) at 10 kHz.

Solution

(a) At 100 Hz

$$X_L = 2\pi \times 100 \times 10 \times 10^{-3}$$

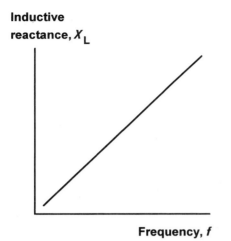

Figure 4.8 Variation of reactance with frequency for an inductor

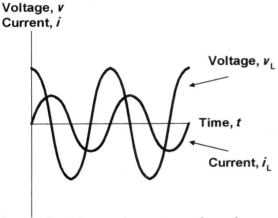

Figure 4.9 Voltage and current waveforms for a pure inductor (the voltage leads the current by 90°)

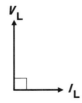

Figure 4.10 Phasor diagram for a pure inductor

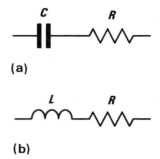

Figure 4.11 (a) C and R in series (this circuit exhibits **impedance**); (b) L and R in series (this circuit exhibits **impedance**)

Thus

$$X_L = 6.28 \ \Omega$$

(b) At 10 kHz

$$X_L = 2\pi \times 10\ 000 \times 10 \times 10^{-3}$$

Thus

$$X_L = 628 \ \Omega$$

Example 4.10

A 100 mH inductor of negligible resistance is to form part of a filter which carries a current of 20 mA at 400 Hz. What voltage drop will be developed across the inductor?

Solution

The reactance of the inductor will be given by:

$$X_L = 2\pi \times 400 \times 100 \times 10^{-3} = 251 \ \Omega$$

The r.m.s. voltage developed across the inductor will be given by:

$$V_L = I_L \times X_L = 20 \text{ mA} \times 251 \ \Omega = 5.02 \text{ V}$$

In this example, it is important to note that we have assumed that the d.c. resistance of the inductor is negligible by comparison with its reactance. Where this is not the case, it is necessary to determine the impedance of the component and use this to determine the voltage drop.

Impedance

Figure 4.11 shows two circuits which contain both resistance and reactance. These circuits are said to exhibit impedance (a combination of resistance and

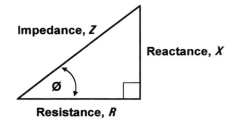

Figure 4.12   The impedance triangle

reactance) which, like resistance and reactance, is measured in ohms. The impedance of the circuits shown in Fig. 4.11 is simply the ratio of supply voltage (V_S) to supply current (I_S).

The impedance of the simple C–R and L–R circuits shown in Fig. 4.11 can be found by using the impedance triangle shown in Fig. 4.12.

In either case, the impedance of the circuit is given by:

$$Z = \sqrt{(X^2 + R^2)}$$

and the phase angle (between V_S and I_S) is given by:

$$\phi = \tan^{-1}(X/R)$$

where Z is the impedance (in ohms), X is the reactance, either capacitive or inductive (expressed in ohms), R is the resistance (in ohms), and ϕ is the phase angle in radians.

Example 4.11

A 2 µF capacitor is connected in series with a 100 Ω resistor across a 115 V 400 Hz a.c. supply. Determine the impedance of the circuit and the current taken from the supply.

Solution

First we must find the reactance of the capacitor, X_C:

$$X_C = \frac{1}{2\pi f C} = \frac{1}{6.28 \times 400 \times 2 \times 10^{-6}}$$

$$= \frac{10^6}{5024} = 199 \; \Omega$$

Now we can find the impedance of the C–R series circuit:

$$Z = \sqrt{(X_C^2 + R^2)} = \sqrt{(199^2 + 100^2)} = \sqrt{49\,601}$$
$$= 223 \; \Omega$$

The current taken from the supply can now be found:

$$I_S = V_S/Z = 115/223 = 0.52 \; A$$

Power factor

The power factor in an a.c. circuit containing resistance and reactance is simply the ratio of true power to apparent power. Hence:

$$\text{power factor} = \frac{\text{true power}}{\text{apparent power}}$$

The **true power** in an a.c. circuit is the power which is actually dissipated in the resistive component. Thus:

$$\text{true power} = I_S^2 \times R$$

The **apparent power** in an a.c. circuit is the power which is apparently consumed by the circuit and is the product of the supply current and supply voltage (note that this is not the same as the power which is actually dissipated as heat). Hence:

$$\text{apparent power} = I_S \times V_S \; \text{VA}$$

Hence

$$\text{power factor} = \frac{I_S^2 \times R}{I_S \times V_S} = \frac{I_S^2 \times R}{I_S \times (I_S \times Z)} = \frac{R}{Z}$$

From Fig. 4.12, $R/Z = \cos \phi$ thus:

$$\text{power factor} = R/Z = \cos \phi$$

Example 4.12

A choke having an inductance of 150 mH and resistance of 250 Ω is connected to a 115 V 400 Hz a.c. supply. Determine the power factor of the choke and the current taken from the supply.

Solution

First we must find the reactance of the inductor, X_L.

$$X_L = 2\pi f L = 6.28 \times 400 \times 0.15 = 376.8 \; \Omega$$

We can now determine the power factor:

$$\text{power factor} = R/Z = 250/376.8 = 0.663$$

The impedance of the choke (Z) will be given by:

$$Z = (X_L^2 + R^2) = (376.8^2 + 250^2) = 452 \; \Omega$$

Finally, the current taken from the supply will be:

$$I_S = V_S/Z = 115/452 = 0.254 \; A$$

Figure 4.13   Series resonant $L–C$ circuit

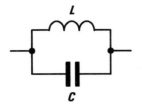

Figure 4.14   Parallel resonant $L–C$ circuit

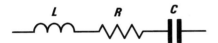

Figure 4.15   Series resonant $L–C–R$ circuit

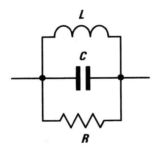

Figure 4.16   Parallel resonant $L–C–R$ circuit

$L–C$ circuits

Two forms of $L–C$ circuits are illustrated in Figs 4.13 and 4.14. Figure 4.13 is a **series resonant** circuit while Fig. 4.14 constitutes a **parallel resonant** circuit. The impedance of both circuits varies in a complex manner with frequency.

The impedance of the series circuit in Fig. 4.13 is given by:

$$Z = \sqrt{(X_L - X_C)^2}$$

where Z is the impedance of the circuit (in ohms), and X_L and X_C are the reactances of the inductor and capacitor respectively (both expressed in ohms).

The phase angle (between the supply voltage and current) will be $+\pi/2$ rad (i.e. $+90°$) when $X_L > X_C$ (above resonance) or $-\pi/2$ rad (or $-90°$) when $X_C > X_L$ (below resonance).

At a particular frequency (known as the **series resonant frequency**) the reactance of the capacitor (X_C) will be equal in magnitude (but of opposite sign) to that of the inductor (X_L). The impedance of the circuit will thus be zero at resonance. The supply current will have a maximum value at resonance (infinite in the case of a perfect series resonant circuit supplied from an ideal voltage source!).

The impedance of the parallel circuit in Fig. 4.14 is given by:

$$Z = \frac{X_L \times X_C}{\sqrt{(X_L - X_C)^2}}$$

where Z is the impedance of the circuit (in Ω), and X_L and X_C are the reactances of the inductor and capacitor, respectively (both expressed in Ω).

The phase angle (between the supply voltage and current) will be $+\pi/2$ rad (i.e. $+90°$) when $X_L > X_C$ (above resonance) or $-\pi/2$ rad (or $-90°$) when $X_C > X_L$ (below resonance).

At a particular frequency (known as the **parallel resonant frequency**) the reactance of the capacitor (X_C) will be equal in magnitude (but of opposite sign) to that of the inductor (X_L). At resonance, the denominator in the formula for impedance becomes zero and thus the circuit has an infinite impedance at resonance. The supply current will have a minimum value at resonance (zero in the case of a perfect parallel resonant circuit).

$L–C–R$ networks

Two forms of $L–C–R$ network are illustrated in Figs 4.15 and 4.16; Fig. 4.15 is series resonant while Fig. 4.16 is parallel resonant. As in the case of their simpler $L–C$ counterparts, the impedance of each circuit varies in a complex manner with frequency.

The impedance of the series circuit of Fig. 4.15 is given by:

$$Z = \sqrt{R^2 + (X_L - X_C)^2}$$

where Z is the impedance of the series circuit (in ohms), R is the resistance (in ohms), X_L is the inductive reactance (in ohms) and X_C is the capacitive reactance (also in ohms). At resonance the circuit has a minimum impedance (equal to R).

The phase angle (between the supply voltage and current) will be given by:

$$\phi = \tan^{-1}\frac{X_L - X_C}{R}$$

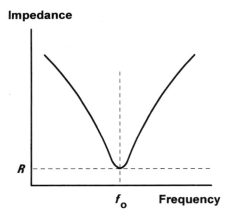

Figure 4.17   Impedance versus frequency for a series *L–C–R* **acceptor circuit**

The impedance of the parallel circuit of Fig. 4.16 is given by:

$$Z = \frac{R \times X_L \times X_C}{\sqrt{(X_L^2 \times X_C^2) + R^2(X_L - X_C)^2}}$$

where Z is the impedance of the parallel circuit (in ohms), R is the resistance (in ohms), X_L is the inductive reactance (in ohms) and X_C is the capacitive reactance (also in ohms). At resonance the circuit has a maximum impedance (equal to R).

The phase angle (between the supply voltage and current) will be given by:

$$\phi = \tan^{-1}\frac{R(X_C - X_L)}{X_L \times X_C}$$

Resonance

The frequency at which the impedance is minimum for a series resonant circuit or maximum in the case of a parallel resonant circuit is known as the resonant frequency. The resonant frequency is given by:

$$f_o = \frac{1}{2\pi\sqrt{LC}}\quad \text{Hz}$$

where f_o is the resonant frequency (in hertz), L is the inductance (in henries) and C is the capacitance (in farads).

Typical impedance–frequency characteristics for series and parallel tuned circuits are shown in Figs 4.17 and 4.18. The series *L–C–R* tuned circuit has a minimum impedance at resonance (equal to R) and thus maximum current will flow. The circuit is

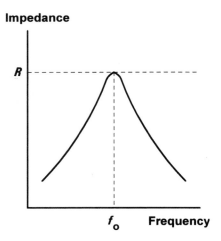

Figure 4.18   Impedance versus frequency for a parallel *L–C–R* **rejector circuit**

consequently known as an **acceptor circuit**. The parallel *L–C–R* tuned circuit has a maximum impedance at resonance (equal to R) and thus minimum current will flow. The circuit is consequently known as a **rejector circuit**.

Quality factor

The quality of a resonant (or tuned) circuit is measured by its **Q-factor**. The higher the Q-factor, the sharper the response (narrower bandwidth), conversely the lower the Q-factor, the flatter the response (wider bandwidth), see Fig. 4.19. In the case of the series tuned circuit, the Q-factor will increase as the resistance, R, decreases. In the case of the parallel tuned circuit, the Q-factor will increase as the resistance, R, increases.

The response of a tuned circuit can be modified by incorporating a resistance of appropriate value either to 'dampen' or 'sharpen' the response.

The relationship between bandwidth and Q-factor is:

$$\text{bandwidth} = f_2 - f_1 = \frac{f_o}{Q}\quad \text{Hz}$$

Example 4.13

A parallel *L–C* circuit is to be resonant at a frequency of 400 Hz. If a 100 mH inductor is available, determine the value of capacitance required.

Impedance

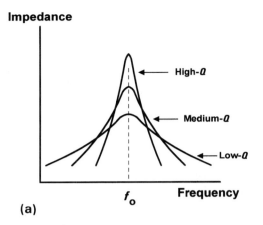

(a)

Voltage

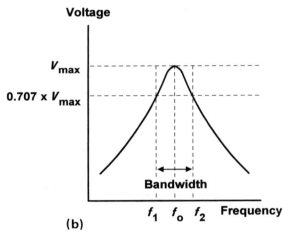

(b)

Figure 4.19 (a) Effect of Q-factor on the response of a parallel resonant circuit (the response is similar, but inverted, for a series resonant circuit); (b) bandwidth

Solution

Re-arranging the formula $f_o = 1/2\pi\sqrt{LC}$ to make C the subject gives:

$$C = \frac{1}{f_o{}^2(2\pi)^2 L}$$

Thus

$$C = \frac{1}{400^2 \times 39.4 \times 100 \times 10^{-3}} \text{ F}$$

or

$$C = \frac{1}{160 \times 10^3 \times 39.4 \times 100 \times 10^{-3}} \text{ F}$$

Hence $C = 1.58$ μF.

This value can be realized from preferred values using a 2.2 μF capacitor connected in series with a 5.6 μF capacitor.

Example 4.14

A series L–C–R circuit comprises an inductor of 20 mH, a capacitor of 10 nF, and a resistor of 100 Ω. If the circuit is supplied with a sinusoidal signal of 1.5 V at a frequency of 2 kHz, determine the current supplied and the voltage developed across the resistor.

Solution

First we need to determine the values of inductive reactance (X_L) and capacitive reactance (X_C):

$$X_L = 2\pi f L = 6.28 \times 2 \times 10^3 \times 20 \times 10^{-3}$$

Thus $X_L = 251.2$ Ω.

$$X_C = \frac{1}{2\pi f C} = \frac{1}{6.28 \times 2 \times 10^3 \times 100 \times 10^{-9}}$$

Thus $X_C = 796.2$ Ω.

The impedance of the series circuit can now be calculated:

$$Z = \sqrt{R^2 + (X_L - X_C)^2}$$
$$= \sqrt{100^2 + (251.2 - 796.2)^2}$$

thus

$$Z = \sqrt{10\,000 + 297\,025} = \sqrt{307\,025} = 554 \ \Omega$$

The current flowing in the series circuit will be given by:

$$I = V/Z = 1.5/554 = 2.7 \text{ mA}$$

The voltage developed across the resistor can now be calculated using:

$$V = IR = 2.7 \text{ mA} \times 100 \ \Omega = 270 \text{ mV}$$

Transformers

Transformers provide us with a means of coupling a.c. power or signals from one circuit to another. Voltage may be **stepped-up** (secondary voltage greater than primary voltage) or **stepped-down** (secondary voltage less than primary voltage). Since no increase in power is possible (transformers are passive components like resistors, capacitors and inductors) an increase in secondary voltage can only

Table 4.2   Characteristics of common types of transformer

Parameter	Transformer type		
	Ferrite cored	Iron cored	Iron cored
Typical power rating	Less than 10 W	100 mW to 50 W	3 VA to 500 VA
Typical regulation	(see note)	(see note)	5% to 15%
Typical frequency range (Hz)	1 k to 10 M	50 to 20 k	45 to 400
Typical applications	Pulse circuits, RF power amplifiers	AF amplifiers	Power supplies

Note: Usually unimportant for this type of transformer.

be achieved at the expense of a corresponding reduction in secondary current, and vice versa (in fact, the secondary power will be very slightly less than the primary power due to losses within the transformer). Typical applications for transformers include stepping-up or stepping-down mains voltages in power supplies, coupling signals in AF amplifiers to achieve impedance matching and to isolate d.c. potentials associated with active components. The electrical characteristics of a transformer are determined by a number of factors including the core material and physical dimensions.

The specifications for a transformer usually include the rated primary and secondary voltages and currents the required power rating (i.e. the maximum power, usually expressed in volt–amperes, VA) which can be continuously delivered by the transformer under a given set of conditions), the frequency range for the component (usually stated as upper and lower working frequency limits), and the **regulation** of a transformer (usually expressed as a percentage of full-load). This last specification is a measure of the ability of a transformer to maintain its rated output voltage under load.

Table 4.1 summarizes the properties of three common types of transformer. Figure 4.20 shows the construction of a typical iron-cored power transformer.

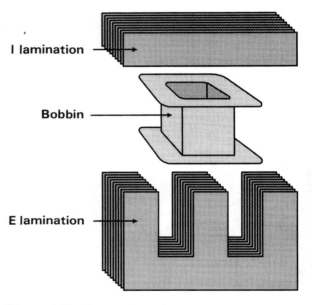

Figure 4.20   Construction of a typical iron-cored transformer

Voltage and turns ratio

The principle of the transformer is illustrated in Fig. 4.21. The primary and secondary windings are wound on a common low-reluctance magnetic core. The alternating flux generated by the primary winding is therefore coupled into the secondary winding (very little flux escapes due to leakage). A sinsuoidal current flowing in the primary winding produces a sinusoidal flux. At any instant the flux in the transformer is given by the equation:

$$\phi = \phi_{max} \sin(2\pi f t)$$

where ϕ_{max} is the maximum value of flux (in Webers), f is the frequency of the applied current (in hertz), and t is the time in seconds.

The r.m.s. value of the primary voltage (V_P) is given by:

$$V_P = 4.44 f N_P \phi_{max}$$

Similarly, the r.m.s. value of the secondary voltage (V_S) is given by:

$$V_S = 4.44 f N_S \phi_{max}$$

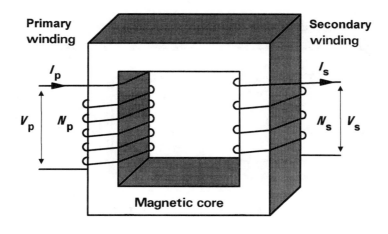

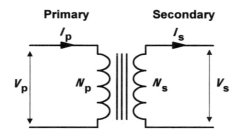

Figure 4.21 The transformer principle

Now

$$V_P/V_S = N_P/N_S$$

where N_P/N_S is the **turns ratio** of the transformer.

Assuming that the transformer is loss-free, primary and secondary powers (P_P and P_S, respectively) will be identical. Hence:

$$P_P = P_S \text{ thus } V_P \times I_P = V_S \times I_S$$

Hence

$$V_P/V_S = I_S/I_P \text{ and } I_S/I_P = N_P/N_S$$

Finally, it is sometimes convenient to refer to a **turns-per-volt** rating for a transformer. This rating is given by:

$$\text{t.p.v.} = N_P/V_P = N_S/V_S$$

Example 4.15

A transformer has 2000 primary turns and 120 secondary turns. If the primary is connected to a 220 V r.m.s. a.c. mains supply, determine the secondary voltage.

Solution

Since $V_P/V_S = N_P/N_S$,

$$V_S = \frac{N_S \times V_P}{N_P} = \frac{120 \times 220}{2000} = 13.2 \text{ V}$$

Example 4.16

A transformer has 1200 primary turns and is designed to operated with a 200 V a.c. supply. If the transformer is required to produce an output of 10 V, determine the number of secondary turns required.

Solution

Since $V_P/V_S = N_P/N_S$,

$$N_S = \frac{N_P \times V_S}{V_P} = \frac{1200 \times 10}{200} = 60 \text{ turns}$$

Circuit symbols introduced in this chapter

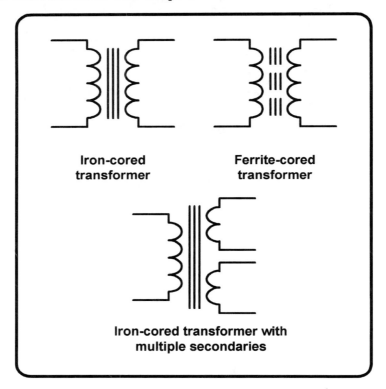

Iron-cored
transformer

Ferrite-cored
transformer

Iron-cored transformer with
multiple secondaries

Figure 4.22

Formulae introduced in this chapter

Sine wave voltage:
(page 69)

$v = V_{max} \sin(2\pi ft)$

Frequency and periodic time:
(page 70)

$t = 1/f$ and $f = 1/t$

Peak and r.m.s. values for a sine wave:
(page 70)

$V_{pk} = 1.414 V_{r.m.s.}$
and $V_{r.m.s.} = 0.707 V_{pk}$

Inductive reactance:
(page 72)

$X_L = 2\pi fL$

Capacitive reactance:
(page 71)

$X_C = \dfrac{1}{2\pi fC}$

Impedance of C–R or L–R in series:
(page 74)

$Z = \sqrt{(X^2 + R^2)}$

Phase angle for C–R or L–R in series:
(page 74)

$\phi = \tan^{-1} (X/R)$

Power factor:
(page 74)

$\text{P.F.} = \dfrac{\text{true power}}{\text{apparent power}}$

$\text{P.F.} = \cos \phi = R/Z$

Resonant frequency of a series resonant tuned circuit:
(page 76)

$f_o = \dfrac{1}{2\pi\sqrt{LC}}$

Bandwidth of a tuned circuit:
(page 76)

$$B/W = f_2 - f_1 = \frac{f_o}{Q}$$

Q-factor for a series tuned circuit:
(page 76)

$$Q = \frac{2\pi f L}{R}$$

Flux in a transformer:
(page 78)

$$\phi = \phi_{max} \sin(2\pi f t)$$

Transformer voltages:
(page 78)

$$V_P = 4.44 f N_P \phi_{max}$$
$$V_S = 4.44 f N_S \phi_{max}$$

Voltage and turns ratio:
(page 79)

$$V_P/V_S = N_P/N_S$$

Current and turns ratio:
(page 79)

$$I_S/I_P = N_P/N_S$$

Turns-per-volt:
(page 79)

$$N_P/V_P = N_S/V_S$$

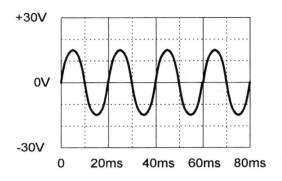

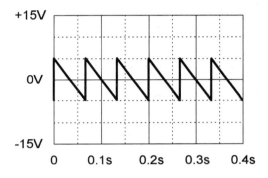

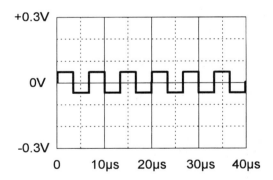

Figure 4.23

Problems

4.1 A sine wave has a frequency of 250 Hz and an amplitude of 50 V. Determine its periodic time and r.m.s. value.

4.2 A sinusoidal voltage has an r.m.s. value of 240 V and a period of 16.7 ms. What is the frequency and peak value of the voltage?

4.3 Determine the frequency and peak–peak values of each of the waveforms shown in Fig. 4.23.

4.4 A sine wave has a frequency of 100 Hz and an amplitude of 20 V. Determine the instantaneous value of voltage (a) 2 ms and (b) 9 ms from the start of a cycle.

4.5 A sinusoidal current of 20 mA pk–pk flows in a resistor of 1.5 kΩ. Determine the r.m.s. voltage applied.

4.6 Determine the reactance of a 220 nF capacitor at (a) 20 Hz and (b) 5 kHz.

4.7 A 47 nF capacitor is connected across the 240 V 50 Hz mains supply. Determine the r.m.s. current flowing in the capacitor.

4.8 Determine the reactance of a 33 mH inductor at (a) 50 Hz and (b) 7 kHz.

4.9 A 10 mH inductor of negligible resistance is used to form part of a filter connected in series with a 50 Hz mains supply. What voltage drop will appear across the inductor when a current of 1.5 A is flowing?

4.10   A 10 µF capacitor is connected in series with a 500 Ω resistor across a 110 V 50 Hz a.c. supply. Determine the impedance of the circuit and the current taken from the supply.

4.11   A choke having an inductance of 1 H and resistance of 250 Ω is connected to a 220 V 60 Hz a.c. supply. Determine the power factor of the choke and the current taken from the supply.

4.12   A series-tuned L–C network is to be resonant at a frequency of 1.8 kHz. If a 60 mH inductor is available, determine the value of capacitance required.

4.13   A parallel resonant circuit employs a fixed inductor of 22 µH and a variable tuning capacitor. If the maximum and minimum values of capacitance are respectively 20 pF and 365 pF, determine the effective tuning range for the circuit.

4.14   A series L–C–R circuit comprises an inductor of 15 mH (with negligible resistance), a capacitor of 220 nF and a resistor of 100 Ω. If the circuit is supplied with a sinusoidal signal of 1.5 V at a frequency of 2 kHz, determine the current supplied and the voltage developed across the capacitor.

4.15   A 470 µH inductor has a resistance of 20 Ω. If the inductor is connected in series with a capacitor of 680 pF, determine the resonant frequency, Q-factor, and bandwidth of the circuit.

4.16   A transformer has 1600 primary turns and 120 secondary turns. If the primary is connected to a 240 V r.m.s. a.c. mains supply, determine the secondary voltage.

4.17   A transformer has 800 primary turns and 60 secondary turns. If the secondary is connected to a load resistance of 15 Ω, determine the value of primary voltage required to produce a power of 22.5 W in the load (assume that the transformer is loss-free).

(Answers to these problems appear on page 202.)

5

Semiconductors

This chapter introduces devices that are made from materials that are neither conductors nor insulators. These **semiconductor** materials form the basis of diodes, thyristors, triacs, transistors and integrated circuits. We start this chapter with a brief introduction to the principles of semiconductors before going on to examine the characteristics of each of the most common types of semiconductor.

In Chapter 1 we described the simplified structure of an atom and showed that it contains both negative charge carriers (electrons) and positive charge carriers (protons). Electrons each carry a single unit of negative electric charge while protons each exhibit a single unit of positive charge. Since atoms normally contain an equal number of electrons and protons, the net charge present will be zero. For example, if an atom has eleven electrons, it will also contain eleven protons. The end result is that the negative charge of the electrons will be exactly balanced by the positive charge of the protons.

Electrons are in constant motion as they orbit around the nucleus of the atom. Electron orbits are organized into shells. The maximum number of electrons present in the first shell is 2, in the second shell 8, and in the third, fourth and fifth shells it is 18, 32 and 50, respectively. In electronics, only the electron shell furthermost from the nucleus of an atom is important. It is important to note that the movement of electrons only involves those present in the outer **valence shell**.

If the valence shell contains the maximum number of electrons possible the electrons are rigidly bonded together and the material has the properties of an insulator. If, however, the valence shell does not have its full complement of electrons, the electrons can be easily loosened from their orbital bonds, and the material has the properties associated with an electrical conductor.

An isolated silicon atom contains four electrons in its valence shell. When silicon atoms combine to form a solid crystal, each atom positions itself between four other silicon atoms in such a way that the valence shells overlap from one atom to another. This causes each individual valence electron to be shared by two atoms, as shown in Fig. 5.1. By sharing the electrons between four adjacent atoms, each individual silicon atom *appears* to have eight electrons in its valence shell. This sharing of valence electrons is called **covalent bonding**.

In its pure state, silicon is an insulator because the covalent bonding rigidly holds all of the electrons leaving no free (easily loosened) electrons to conduct current. If, however, an atom of a different element (i.e. an **impurity**) is introduced that has five electrons in its valence shell, a surplus electron will be present. These **free electrons** become available for use as **charge carriers** and they can be made to move through the lattice by applying an external potential difference to the material.

Similarly, if the impurity element introduced into the pure silicon lattice has three electrons in its valence shell, the absence of the fourth electron needed for proper covalent bonding will produce a

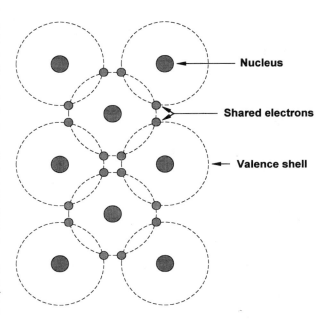

Figure 5.1 Lattice showing covalent bonding

number of spaces into which electrons can fit. These spaces are referred to as **holes**. Once again, current will flow when an external potential difference is applied to the material.

Regardless of whether the impurity element produces surplus electrons or holes, the material will no longer behave as an insulator, neither will it have the properties that we normally associate with a metallic conductor. Instead, we call the material a **semiconductor** – the term simply indicates that the substance is no longer a good insulator or a good conductor but is somewhere in between!

The process of introducing an atom of another (impurity) element into the lattice of an otherwise pure material is called **doping**. When the pure material is been doped with an impurity with five electrons in its valence shell (i.e. a **pentavalent impurity**) it will become an **N-type** material. If, however, the pure material is doped with an impurity having three electrons in its valence shell (i.e. a **trivalent impurity**) it will become **P-type** material. N-type semiconductor material contains an excess of negative charge carriers, and P-type material contains an excess of positive charge carriers.

Semiconductor diodes

When a junction is formed between N-type and P-type semiconductor materials, the resulting device is called a diode. This component offers an extremely low resistance to current flow in one direction and an extremely high resistance to current flow in the other. This characteristic allows the diode to be used in applications that require a circuit to behave differently according to the direction of current flowing in it.

An ideal diode would pass an infinite current in one direction and no current at all in the other direction. In addition, the diode would start to conduct current when the smallest of voltages was present. In practice, a small voltage must be applied before conduction takes place. Furthermore a small **leakage current** will flow in the **reverse direction**. This leakage current is usually a very small fraction of the current that flows in the **forward direction**.

If the P-type semiconductor material is made positive relative to the N-type material by an amount greater than its **forward threshold voltage** (about 0.6 V if the material is silicon and 0.2 V if the material is germanium), the diode will freely pass

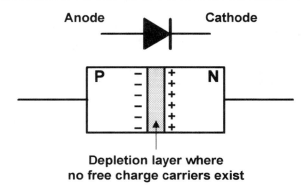

Depletion layer where no free charge carriers exist

Figure 5.2   P–N junction diode

current. If, on the other hand, the P-type material is made negative relative to the N-type material, virtually no current will flow unless the applied voltage exceeds the maximum (breakdown) voltage that the device can withstand. Note that a normal diode will be destroyed if its **reverse breakdown voltage** is exceeded.

A semiconductor junction diode is shown in Fig. 5.2. The connection to the P-type material is referred to as the **anode** while that to the N-type material is called the **cathode**. With no externally applied potential, electrons from the N-type material will cross into the P-type region and fill some of the vacant holes. This action will result in the production of a region either side of the junction in which there are no free charge carriers. This zone is known as the **depletion region**.

Figure 5.3 shows a junction diode in which the anode is made positive with respect to the cathode. In this **forward-biased** condition, the diode freely passes current. Figure 5.4 shows a diode with the cathode made positive with respect to the cathode. In this **reverse-biased** condition, the diode passes a negligible amount of current. In the freely conducting forward-biased state, the diode acts rather like a closed switch. In the reverse-biased state, the diode acts like an open switch.

If a positive voltage is applied to the P-type material, the free positive charge carriers will be repelled and they will move away from the positive potential towards the junction. Likewise, the negative potential applied to the N-type material will cause the free negative charge carriers to move away from the negative potential towards the junction.

When the positive and negative charge carriers arrive at the junction, they will attract one another and combine (recall that unlike charges attract). As

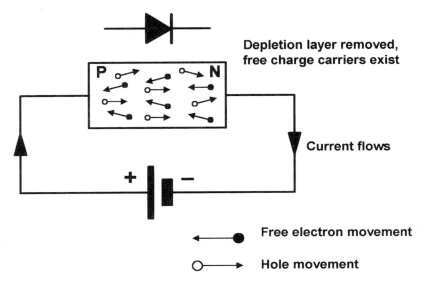

Figure 5.3 Forward biased P–N junction

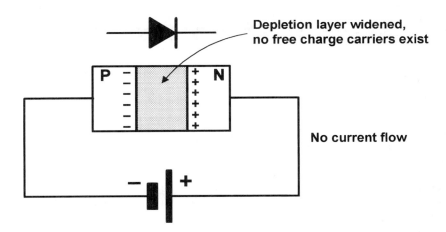

Figure 5.4 Reverse biased P–N junction

each negative and positive charge carrier combine at the junction, a new negative and positive charge carrier will be introduced to the semiconductor material from the voltage source. As these new charge carriers enter the semiconductor material, they will move toward the junction and combine. Thus, current flow is established and it will continue for as long as the voltage is applied.

As stated earlier, the **forward threshold voltage** must be exceeded before the diode will conduct. The forward threshold voltage must be high enough to completely remove the depletion layer and force charge carriers to move across the junction. With silicon diodes, this forward threshold voltage is approximately 0.6 V to 0.7 V. With germanium diodes, the forward threshold voltage is approximately 0.2 V to 0.3 V.

Figure 5.5 shows typical characteristics for small germanium and silicon diodes. It is worth noting that diodes are limited by the amount of forward current and reverse voltage they can withstand. This limit is based on the physical size and construction of the diode.

In the case of a reverse biased diode, the P-type material is negatively biased relative to the N-type material. In this case, the negative potential applied

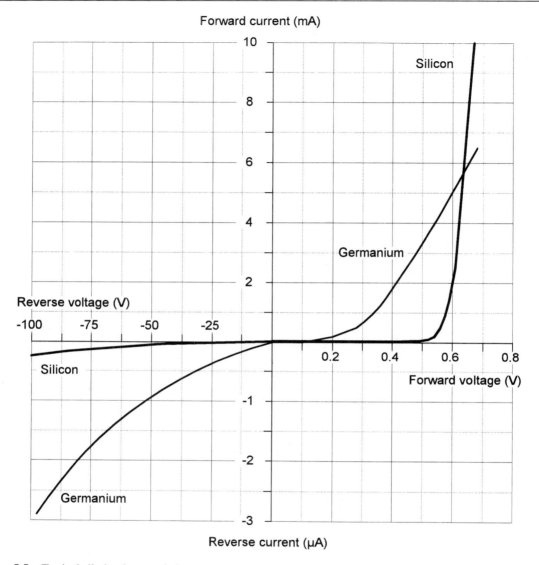

Figure 5.5   Typical diode characteristics

to the P-type material attracts the positive charge carriers, drawing them away from the junction. Likewise, the positive potential applied to the N-type material attracts the negative charge carriers away from the junction. This leaves the junction area depleted; virtually no charge carriers exist. Therefore, the junction area becomes an insulator, and current flow is inhibited.

The reverse bias potential may be increased to the reverse breakdown voltage for which the particular diode is rated. As in the case of the maximum forward current rating, the reverse breakdown voltage is specified by the manufacturer. The reverse breakdown voltage is usually very much higher than the forward threshold voltage. A typical general-purpose diode may be specified as having a forward threshold voltage of 0.6 V and a reverse breakdown voltage of 200 V. If the latter is exceeded, the diode may suffer irreversible damage. It is also worth noting that, where diodes are designed for use as rectifiers, manufacturers often quote **peak inverse voltage** (PIV) or **maximum reverse repetitive voltage** (V_{RRM}) rather than maximum reverse breakdown voltage.

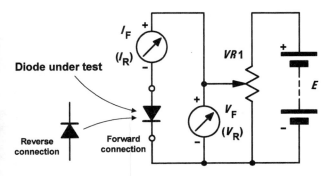

Figure 5.6 Diode test circuit

Figure 5.6 shows a test circuit for obtaining diode characteristics (note that the diode must be reverse connected in order to obtain the reverse characteristic).

Diode types

Diodes are often divided into **signal** or **rectifier** types according to their principal field of application. Signal diodes require consistent forward characteristics with low forward voltage drop. Rectifier diodes need to be able to cope with high values of reverse voltage and large values of forward current, consistency of characteristics is of secondary importance in such applications. Table 5.1 summarizes the characteristics of some common semiconductor diodes.

Example 5.1

The characteristic shown in Fig. 5.9 refers to a germanium diode. Determine the resistance of the diode when (a) $I_F = 2.5$ mA and (b) $V_F = 0.65$ V.

Solution

See Fig. 5.9.

Zener diodes

Zener diodes are heavily doped silicon diodes which, unlike normal diodes, exhibit an abrupt reverse breakdown at relatively low voltages (typically less than 6 V). A similar effect occurs in less heavily doped diodes. These **avalanche diodes** also exhibit a rapid breakdown with negligible current flowing below the avalanche voltage and a relatively large current flowing once the avalanche voltage has been reached. For avalanche diodes, this breakdown voltage usually occurs at voltages above 6 V. In practice, however, both types of diode are referred to as zener diodes. A typical characteristic for a 5.1 V zener diode is shown in Fig. 5.7.

Whereas reverse breakdown is a highly undesirable effect in circuits that use conventional diodes, it can be extremely useful in the case of zener diodes where the breakdown voltage is precisely known. When a diode is undergoing reverse breakdown *and provided its maximum ratings are not exceeded* the voltage appearing across it will remain substantially constant (equal to the nominal zener voltage) regardless of the current flowing. This property makes the zener diode ideal for use as a **voltage regulator** (see Chapter 6).

Zener diodes are available in various families (according to their general characteristics, encapsulation and power ratings) with reverse breakdown (zener) voltages in the E12 and E24 series (ranging from 2.4 V to 91 V). Table 5.2 summarizes the characteristics of common zener diodes, while Fig. 5.10 shows typical encapsulations used for conventional diodes and zener diodes.

Table 5.1 Characteristics of some common semiconductor diodes

Device	Material	PIV	I_F max.	I_R max.	Application
1N4148	Silicon	100 V	75 mA	25 nA	General purpose
1N914	Silicon	100 V	75 mA	25 nA	General purpose
AA113	Germanium	60 V	10 mA	200 μA	RF detector
OA47	Germanium	25 V	110 mA	100 μA	Signal detector
OA91	Germanium	115 V	50 mA	275 μA	General purpose
1N4001	Silicon	50 V	1 A	10 μA	Low-voltage rectifier
1N5404	Silicon	400 V	3 A	10 μA	High-voltage rectifier
BY127	Silicon	1250 V	1 A	10 μA	High-voltage rectifier

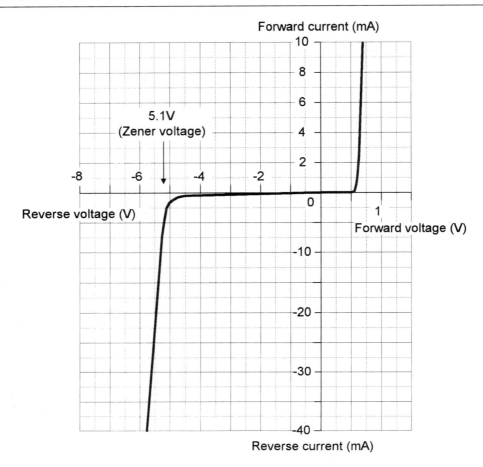

Figure 5.7   Typical characteristics for a 5.1 V zener diode

Table 5.2   Characteristics of some common zener diodes

BZY88 series	Miniature glass encapsulated diodes rated at 500 mW (at 25°C). Zener voltages range from 2.7 V to 15 V (voltages are quoted for 5 mA reverse current at 25°C)
BZX61 series	Encapsulated alloy junction rated at 1.3 W (25°C ambient). Zener voltages range from 7.5 V to 72 V
BZX85 series	Medium-power glass-encapsulated diodes rated at 1.3 W and offering zener voltages in the range 5.1 V to 62 V
BZY93 series	High-power diodes in stud mounting encapsulation. Rated at 20 W for ambient temperatures up to 75°C. Zener voltages range from 9.1 V to 75 V
1N5333 series	Plastic encapsulated diodes rated at 5 W. Zener voltages range from 3.3 V to 24 V

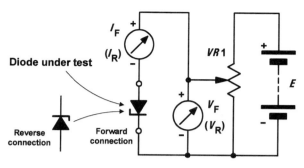

Figure 5.8 Zener diode test circuit

Figure 5.8 shows a test circuit for obtaining zener diode characteristics. The circuit is shown with the diode connected in the forward direction and it must be reverse connected in order to obtain the reverse characteristic.

Diode coding

The European system for classifying semiconductor diodes involves an alphanumeric code which employs either two letters and three figures (general-purpose diodes) or three letters and two figures (special-purpose diodes). Table 5.3 shows how diodes are coded. Typical diode packages and markings (the stripe indicates the cathode connection) are shown in Fig. 5.10.

Example 5.2

Identify each of the following diodes:

(a) AA113
(b) BB105
(c) BZY88C4V7.

Solution

Diode (a) is a general-purpose germanium diode.

Diode (b) is a silicon variable capacitance diode.

Diode (c) is a silicon zener diode having ±5% tolerance and 4.7 V zener voltage.

Variable capacitance diodes

The capacitance of a reverse-biased diode junction will depend on the width of the depletion layer

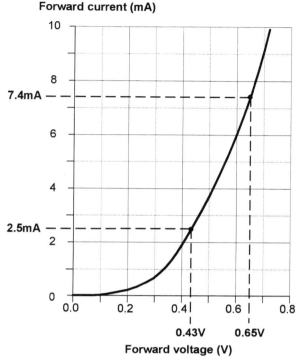

(a) R = 0.43V/2.5mA = 172 ohm
(b) R = 0.65V/7.4mA = 88 ohm

Figure 5.9

which, in turn, varies with the reverse voltage applied to the diode. This allows a diode to be used as a voltage controlled capacitor. Diodes that are specially manufactured to make use of this effect (and which produce comparatively large changes in capacitance for a small change in reverse voltage) are known as variable capacitance diodes (or **varicaps**). Such diodes are used (often in pairs) to provide tuning in radio and TV receivers. A typical characteristic for a variable capacitance diode is shown in Fig. 5.11. Table 5.4 summarizes the characteristics of several common variable capacitance diodes.

Thyristors

Thyristors (or **silicon controlled rectifiers**) are three-terminal devices which can be used for switching and a.c. power control.

Thyristors can switch very rapidly from a conducting to a non-conducting state. In the off state,

Table 5.3   Diode coding

First letter – semiconductor material: A   Germanium
                                          B   Silicon
                                          C   Gallium arsenide, etc.
                                          D   Photodiodes, etc.

Second letter – application: A   General-purpose diode
                               B   Variable-capacitance diode
                               E   Tunnel diode
                               P   Photodiode
                               Q   Light emitting diode
                               T   Controlled rectifier
                               X   Varactor diode
                               Y   Power rectifier
                               Z   Zener diode

Third letter – in the case of diodes for specialized applications, the third letter does not generally have any particular significance

Zener diodes – zener diodes have an additional letter (which appears after the numbers) which denotes the tolerance of the zener voltage. The following letters are used:

A	±1%
B	±2%
C	±5%
D	±10%

Zener diodes also have additional characters which indicate the zener voltage (e.g. 9V1 denotes 9.1 V).

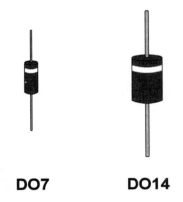

DO7   **DO14**

Figure 5.10   Some common diode packages

the thyristor exhibits negligible leakage current, while in the on state the device exhibits very low resistance. This results in very little power loss within the thyristor even when appreciable power levels are being controlled. Once switched into the conducting state, the thyristor will remain conducting (i.e. it is latched in the on state) until the forward current is removed from the device. In d.c. applications this necessitates the interruption (or disconnection) of the supply before the device can

be reset into its non-conducting state. Where the device is used with an alternating supply, the device will automatically become reset whenever the main supply reverses. The device can then be triggered on the next half-cycle having correct polarity to permit conduction.

Like their conventional silicon diode counterparts, thyristors have anode and cathode connections; control is applied by means of a gate terminal (see Fig. 5.12). The device is triggered into the conducting (on state) by means of the application of a current pulse to this terminal. The effective triggering of a thyristor requires a **gate trigger** pulse having a fast rise time derived from a low-resistance source. Triggering can become erratic when insufficient gate current is available or when the gate current changes slowly. Table 5.5 summarizes the characteristics of several common thyristors.

Triacs

Triacs are a refinement of the thyristor which, when triggered, conduct on both positive and negative half-cycles of the applied voltage. Triacs have three terminals known as main-terminal one (MT1), main

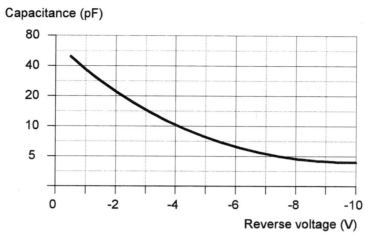

Figure 5.11 Typical characteristics for a variable capacitance diode

Table 5.4 Characteristics of some common variable capacitance diodes

Type	Capacitance	Capacitance ratio	Q-factor
1N5450	33 pF at 4 V	2.6 for 4 V to 60 V	350
MV1404	50 pF at 4 V	>10 for 2 V to 10 V change	200
MV2103	10 pF at 4 V	2 for 4 V to 60 V	400
MV2115	100 pF at 4 V	2.6 for 4 V to 60 V	100

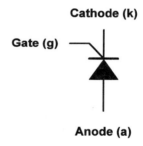

Figure 5.12 Thyristor connections

terminal two (MT2) and gate (G), as shown in Fig. 5.13. Triacs can be triggered by both positive and negative voltages applied between G and MT1 with positive and negative voltages present at MT2 respectively. Triacs thus provide **full-wave control** and offer superior performance in a.c. power control applications when compared with thyristors which only provide **half-wave control**. Table 5.6 summarizes the characteristics of several common triacs.

In order to simplify the design of triggering circuits, triacs are often used in conjunction with diacs (equivalent to a bi-directional zener diode). A typical **diac** conducts heavily when the applied voltage exceeds approximately 30 V in either direction. Once in the conducting state, the resistance of the diac falls to a very low value and thus a relatively large value of current will flow. The characteristic of a typical diac is shown in Fig. 5.14.

Table 5.5 Characteristics of some common thyristors

Type	$I_{F(AV)}$ (A)	V_{RRM} (V)	V_{GT} (V)	I_{GT} (mA)
2N4444	5.1	600	1.5	30
BT106	1	700	3.5	50
BT152	13	600	1	32
BTY79–400R	6.4	400	3	30
TIC106D	3.2	400	1.2	0.200
TIC126D	7.5	400	2.5	20

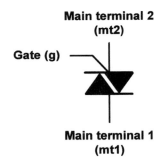

Figure 5.13   Triac connections

Table 5.6   Characteristics of some common triacs

Type	$I_{T(RMS)}$ (A)	V_{RRM} (V)	V_{GT} (V)	$I_{GT(TYP)}$ (mA)
2N6075	4	600	2.5	5
BT139	15	600	1.5	5
TIC206M	4	600	2	5
TIC226M	8	600	2	50

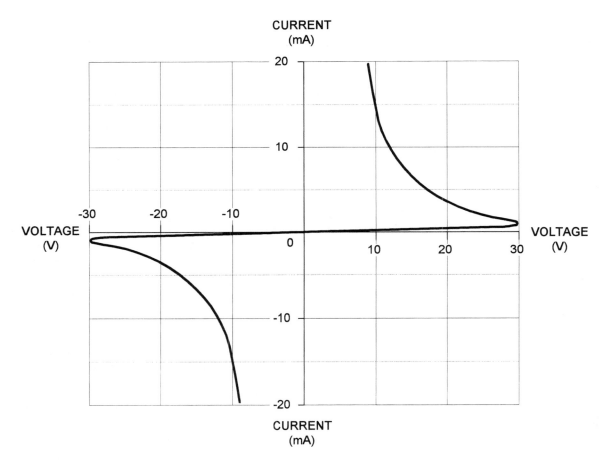

Figure 5.14   Diac characteristics

Table 5.7 Characteristics of some common types of LED

Parameter	Type of LED			
	Standard	Standard	High efficiency	High intensity
Diameter (mm)	3	5	5	5
Max. forward current (mA)	40	30	30	30
Typical forward current (mA)	12	10	7	10
Typical forward voltage drop (V)	2.1	2.0	1.8	2.2
Max. reverse voltage (V)	5	3	5	5
Max. power dissipation (mW)	150	100	27	135
Peak wavelength (nm)	690	635	635	635

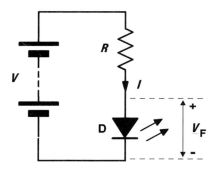

Figure 5.15 Use of a current limiting resistor with an LED

Light emitting diodes

Light emitting diodes (LED) can be used as general-purpose indicators and, compared with conventional filament lamps, operate from significantly smaller voltages and currents. LEDs are also very much more reliable than filament lamps. Most LEDs will provide a reasonable level of light output when a forward current of between 5 mA and 20 mA is applied.

Light emitting diodes are available in various formats with the round types being most popular. Round LEDs are commonly available in the 3 mm and 5 mm (0.2 inch) diameter plastic packages and also in a 5 mm × 2 mm rectangular format. The viewing angle for round LEDs tends to be in the region of 20° to 40°, whereas for rectangular types this is increased to around 100°. Table 5.7 summarizes the characteristics of several common types of LED.

In order to limit the forward current of an LED to an appropriate value, it is usually necessary to include a fixed resistor in series with an LED indicator, as shown in Fig. 5.15. The value of the resistor may be calculated from:

$$R = \frac{V - V_F}{I}$$

where V_F is the forward voltage drop produced by the LED and V is the applied voltage. Note that it is usually safe to assume that V will be 2 V and choose the nearest preferred value for R.

Example 5.3

An LED is to be used to indicate the presence of a 21 V d.c. supply rail. If the LED has a nominal forward voltage of 2.2 V, and is rated at a current of 15 mA, determine the value of series resistor required.

Solution

Here we can use the formula:

$$R = \frac{V - V_F}{I} = \frac{21 \text{ V} - 2.2 \text{ V}}{15 \text{ mA}} = \frac{18.8 \text{ V}}{15 \text{ mA}} = 1.25 \text{ k}\Omega$$

The nearest preferred value is 1.2 kΩ. The power dissipated in the resistor will be given by:

$$P = I \times V = 15 \text{ mA} \times 18.8 \text{ V} = 280 \text{ mW}$$

Hence the resistor should be rated at 0.33 W, or greater.

Bipolar transistors

Transistor is short for **transfer resistor**, a term which provides something of a clue as to how the

Table 5.8   Classes of transistor

Low-frequency	Transistors designed specifically for audio frequency applications (below 100 kHz)
High-frequency	Transistors designed specifically for radio frequency applications (100 kHz and above)
Power	Transistors that operate at significant power levels (such devices are often sub-divided into audio frequency and radio frequency power types)
Switching	Transistors designed for switching applications
Low-noise	Transistors that have low-noise characteristics and which are intended primarily for the amplification of low-amplitude signals
High-voltage	Transistors designed specifically to handle high voltages
Driver	Transistors that operate at medium power and voltage levels and which are often used to precede a final (power) stage which operates at an appreciable power level

Table 5.9   Transistor coding

First letter – semiconductor material: A   Germanium
　　　　　　　　　　　　　　　　　　 B   Silicon

Second letter – application: C   Low-power, low-frequency
　　　　　　　　　　　　　　 D   High-power, low-frequency
　　　　　　　　　　　　　　 F   Low-power, high-frequency
　　　　　　　　　　　　　　 L   High-power, high-frequency

Third letter – in the case of transistors for specialized and industrial applications, the third letter does not generally have any particular significance

device operates; the current flowing in the output circuit is determined by the current flowing in the input circuit. Since transistors are three-terminal devices, one electrode must remain common to both the input and the output.

Transistors fall into two main categories (bipolar and field-effect) and are also classified according to the semiconductor material employed (silicon or germanium) and to their field of application (e.g. general purpose, switching, high-frequency, etc.). Various classes of transistor are available according to the application concerned (see Table 5.8).

Transistor coding

The European system for classifying transistors involves an alphanumeric code which employs either two letters and three figures (general-purpose transistors) or three letters and two figures (special-purpose transistors). Table 5.9 shows how transistors are coded.

Example 5.4

Identify each of the following transistors:

(a) AF115
(b) BC109
(c) BD135
(d) BFY51.

Solution

Transistor (a) is a general-purpose, low-power, high-frequency germanium transistor.

Transistor (b) is a general-purpose, low-power, low-frequency silicon transistor.

Transistor (c) is a general-purpose, high-power, low-frequency silicon transistor.

Transistor (d) is a special-purpose, low-power, high-frequency silicon transistor.

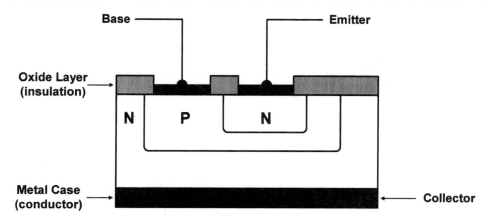

Figure 5.16   NPN transistor construction

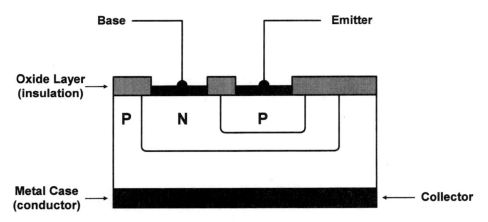

Figure 5.17   PNP transistor construction

Transistor operation

Bipolar transistors generally comprise NPN or PNP junctions of either **silicon** (Si) or **germanium** (Ge) material (see Figs 5.16 and 5.17). The junctions are, in fact, produced in a single slice of silicon by diffusing impurities though a photographically reduced mask. Silicon transistors are superior when compared with germanium transistors in the vast majority of applications (particularly at high temperatures) and thus germanium devices are very rarely encountered.

Figures 5.18 and 5.19, respectively, show a simplified representation of NPN and PNP transistors together with their circuit symbols. In either case the electrodes are labelled **collector**, **base** and **emitter**. Note that each junction within the transistor, whether it be collector–base or base–emitter, constitutes a P–N junction.

Figures 5.20 and 5.21, respectively, show the normal bias voltages applied to NPN and PNP transistors. Note that the base–emitter junction is forward biased and the collector–base junction is reverse biased. The base region is, however, made very narrow so that carriers are swept across it from emitter to collector and only a relatively small current flows in the base. To put this into context, the current flowing in the emitter circuit is typically 100 times greater than that flowing in the base. The direction of conventional current flow is from emitter to collector in the case of a PNP transistor, and collector to emitter in the case of an NPN device.

The equation which relates current flow in the collector, base, and emitter currents is:

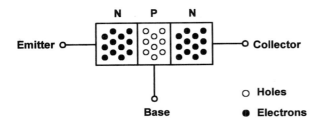

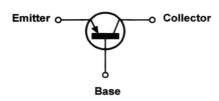

Figure 5.18   Simplified model of an NPN transistor

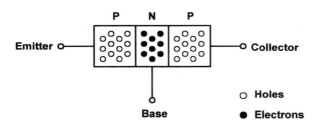

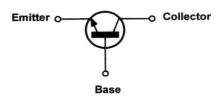

Figure 5.19   Simplified model of a PNP transistor

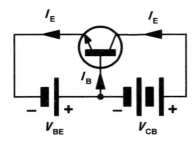

Figure 5.20   Bias voltages and current flow in an NPN transistor

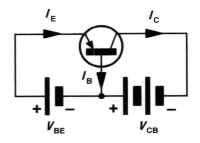

Figure 5.21   Bias voltages and current flow in a PNP transistor

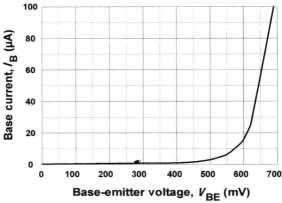

Figure 5.22   Typical input characteristic for a small-signal NPN transistor operating in common-emitter mode

$$I_E = I_B + I_C$$

where I_E is the emitter current, I_B is the base current, and I_C is the collector current (all expressed in the same units).

Bipolar transistor characteristics

The characteristics of a transistor are often presented in the form of a set of graphs relating voltage and current present at the transistor's terminals.

A typical **input characteristic** (I_B plotted against V_{BE}) for a small-signal general-purpose NPN transistor operating in common-emitter mode (see Chapter 7) is shown in Fig. 5.22. This characteristic shows that very little base current flows until the base–emitter voltage (V_{BE}) exceeds 0.6 V. Thereafter, the base current increases rapidly (this

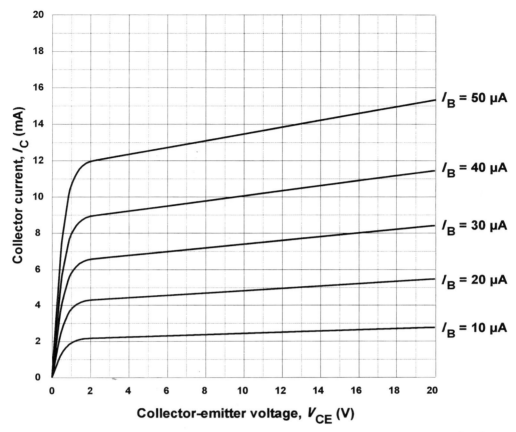

Figure 5.23 Typical family of output characteristics for a small-signal NPN transistor operating in common-emitter mode

characteristic bears a close resemblance to the forward part of the characteristic for a silicon diode, see Fig. 5.5).

Figure 5.23 shows a typical **output characteristic** (I_C plotted against V_{CE}) for a small-signal general-purpose NPN transistor operating in common-emitter mode (see Chapter 7). This characteristic comprises a family of curves, each relating to a different value of base current (I_B). It is worth taking a little time to get familiar with this characteristic as we shall be putting it to good use in Chapter 7. In particular it is important to note the 'knee' that occurs at values of V_{CE} of about 2 V. Also, note how the curves become flattened above this value with the collector current (I_C) not changing very greatly for a comparatively large change in collector–emitter voltage (V_{CE}).

Finally, a typical **transfer characteristic** (I_C plotted against I_B) for a small-signal general-purpose NPN transistor operating in common-emitter mode (see Chapter 7) is shown in Fig. 5.24. This charac-

teristic shows an almost linear relationship between collector current and base current (i.e. doubling the value of base current produces double the value of collector current, and so on). This characteristic is reasonably independent of the value of collector–emitter voltage (V_{CE}) and thus only a single curve is used.

Current gain

The current gain offered by a transistor is a measure of its effectiveness as an amplifying device. The most commonly quoted parameter is that which relates to **common-emitter mode**. In this mode, the input current is applied to the base and the output current appears in the collector (the emitter is effectively common to both the input and output circuits).

The common-emitter current gain is given by:

$$h_{FE} = I_C/I_B$$

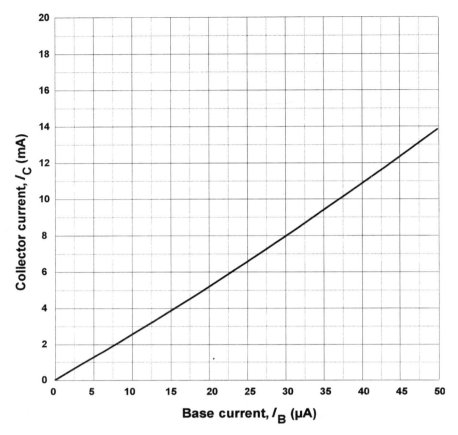

Figure 5.24   Typical transfer characteristic for a small-signal NPN transistor operating in common-emitter mode

where h_{FE} is the **hybrid parameter** which represents **large signal** (d.c.) **forward current gain**, I_C is the collector current, and I_B is the base current. When small (rather than large) signal operation is considered, the values of I_C and I_B are incremental (i.e. small changes rather than static values). The current gain is then given by:

$$h_{fe} = I_c/I_b$$

where h_{fe} is the **hybrid parameter** which represents **small signal** (a.c.) **forward current gain**, I_c is the change in collector current which results from a corresponding change in base current, I_b.

Values of h_{FE} and h_{fe} can be obtained from the transfer characteristic (I_C plotted against I_B) as shown in Figs 5.25 and 5.26. Note that h_{FE} is found from corresponding static values while h_{fe} is found by measuring the slope of the graph. Also note that, if the transfer characteristic is linear, there is little (if any) difference between h_{FE} and h_{fe}.

It is worth noting that current gain (h_{fe}) varies with collector current. For most small-signal transistors, h_{fe} is a maximum at a collector current in the range 1 mA and 10 mA. Current gain also falls to very low values for power transistors when operating at very high values of collector current. Furthermore, most transistor parameters (particularly common-emitter current gain, h_{fe}) are liable to wide variation from one device to the next. It is, therefore, important to design circuits on the basis of the minimum value for h_{fe} in order to ensure successful operation with a variety of different devices.

Transistor parameters are listed in Table 5.10, while Table 5.11 shows the characteristics of several common types of bipolar transistor. Finally, Fig. 5.27 shows a test circuit for obtaining NPN transistor characteristics (the arrangement for a PNP transistor is similar but all meters and supplies must be reversed).

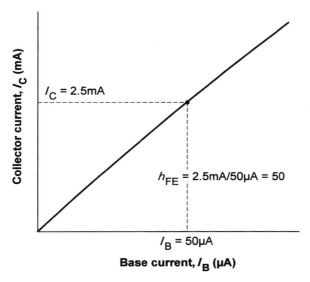

Figure 5.25 Determining the static value of current gain (h_{FE}) from the transfer characteristic

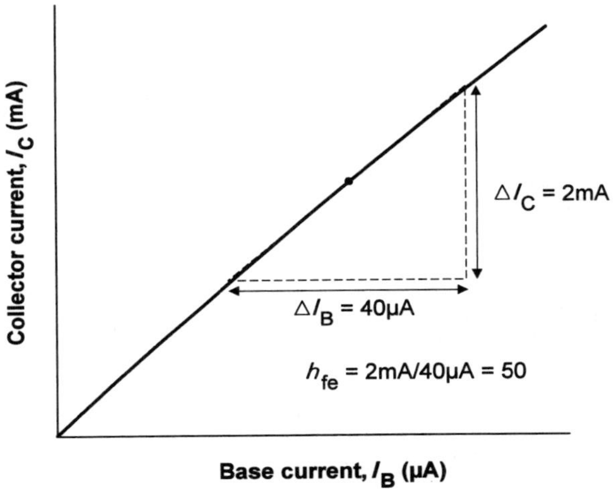

Figure 5.26 Determining the small-signal value of current gain (h_{fe}) from the transfer characteristic

Table 5.10 Bipolar transistor parameters

I_C max.	– the maximum value of collector current
V_{CEO} max.	– the maximum value of collector–emitter voltage with the base terminal left open-circuit
V_{CBO} max.	– the maximum value of collector–base voltage with the base terminal left open-circuit
P_t max.	– the maximum total power dissipation
h_{FE}	– the large signal (static) common-emitter current gain
h_{fe}	– the small-signal common-emitter current gain
h_{fe} max.	– the maximum value of small-signal common-emitter current gain
h_{fe} min.	– the minimum value of small-signal common-emitter current gain
h_{ie}	– the small-signal input resistance (see Chapter 7)
h_{oe}	– the small-signal output conductance (see Chapter 7)
h_{re}	– the small-signal reverse current transfer ratio (see Chapter 7)
f_t typ.	– the transition frequency (i.e. the frequency at which the small-signal common-emitter current gain falls to unity

Example 5.5

A transistor operates with $I_C = 30$ mA and $I_B = 600$ μA. Determine the value of I_E and h_{FE}.

Solution

The value of I_E can be calculated from $I_E = I_C + I_B$, thus $I_E = 30 + 0.6 = 30.6$ mA.

The value of h_{FE} can be calculated from $h_{FE} = I_C/I_B = 30/0.6 = 50$.

Example 5.6

A transistor operates with a collector current of 97 mA and an emitter current of 98 mA. Determine the value of base current and common-emitter current gain.

Table 5.11   Characteristics of some common types of bipolar transistor

Device	Type	I_c max.	V_{ceo} max.	V_{cbo} max.	P_t max.	h_{fe}	at I_c	f_t typ.	Application
BC108	NPN	100 mA	20 V	30 V	300 mW	125	2 mA	250 MHz	General purpose
BCY70	PNP	200 mA	−40 V	−50 V	360 mW	150	2 mA	200 MHz	General purpose
BD131	NPN	3 A	45 V	70 V	15 W	50	250 mA	60 MHz	AF power
BD132	PNP	3 A	−45 V	−45 V	15 W	50	250 mA	60 MHz	AF power
BF180	NPN	20 mA	20 V	20 V	150 mW	100	10 mA	650 MHz	RF amplifier
2N3053	NPN	700 mA	40 V	60 V	800 mW	150	50 mA	100 MHz	Driver
2N3055	NPN	15 A	60 V	100 V	115 W	50	500 mA	1 MHz	LF power
2N3866	NPN	400 mA	30 V	30 V	3 W	105	50 mA	700 MHz	RF driver
2N3904	NPN	200 mA	40 V	60 V	310 mW	150	50 mA	300 MHz	Switching

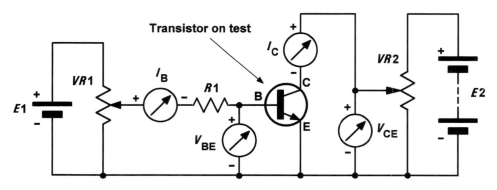

Figure 5.27   NPN transistor test circuit (the arrangement for a PNP transistor is similar but all meters and supplies must be reversed)

Solution

Since $I_E = I_B + I_C$, the base current will be given by:

$$I_B = I_E - I_C = 98 - 97 = 1 \text{ mA}$$

The common-emitter current gain (h_{FE}) will be given by:

$$h_{FE} = I_C/I_B = 97/1 = 97$$

Example 5.7

An NPN transistor is to be used in a regulator circuit in which a collector current of 1.5 A is to be controlled by a base current of 50 mA. What value of h_{FE} will be required? If the device is to be operated with $V_{CE} = 6$ V, which transistor selected from Table 5.11 would be appropriate for this application and why?

Solution

The required current gain can be found from:

$$h_{FE} = I_C/I_B = 1.5 \text{ A}/50 \text{ mA} = 1500 \text{ mA}/50 \text{ mA}$$
$$= 30$$

The most appropriate device would be the BD131. The only other device capable of operating at a collector current of 1.5 A would be a 2N3055. The collector power dissipation will be given by:

$$P_C = I_C \times V_{CE} = 1.5 \text{ A} \times 6 \text{ V} = 9 \text{ W}$$

However, the 2N3055 is rated at 115 W maximum total power dissipation and this is more than ten times the power required.

Example 5.8

A transistor is used in a linear amplifier arrangement. The transistor has small and large signal current gains of 200 and 175, respectively, and bias is arranged so that the static value of collector current is 10 mA. Determine the value of base bias current and the change of output (collector) current that would result from a 10 μA change in input (base) current.

Solution

The value of base bias current can be determined from:

$$I_B = I_C/h_{FE} = 10 \text{ mA}/200 = 50 \text{ μA}$$

The change of collector current resulting from a 10 μA change in input current will be given by:

$$I_c = h_{fe} \times I_b = 175 \times 10 \text{ μA} = 1.75 \text{ mA}$$

Field effect transistors

Field effect transistors (FET) comprise a channel of P-type or N-type material surrounded by material of the opposite polarity. The ends of the channel (in which conduction takes place) form electrodes known as the **source** and **drain**. The effective width of the channel (in which conduction takes place) is controlled by a charge placed on the third (**gate**) electrode. The effective resistance between the source and drain is thus determined by the voltage present at the gate.

Field effect transistors are available in two basic forms; **junction gate** and **insulated gate**. The gate–source junction of a junction gate field effect transistor (JFET) is effectively a reverse-biased P-N junction. The gate connection of an insulated gate field effect transistor (IGFET), on the other hand, is insulated from the channel and charge is capacitively coupled to the channel. To keep things simple, we will consider only JFET devices in this book. Figure 5.28 shows the basic construction of an N-channel JFET.

JFETs offer a very much higher input resistance when compared with bipolar transistors. For example, the input resistance of a bipolar transistor operating in common-emitter mode (see Chapter 7) is usually around 2.5 kΩ. A JFET transistor operating in equivalent common-source mode (see Chapter 7) would typically exhibit an input resistance of 100 MΩ! This feature makes JFET devices ideal for use in applications where a very high input resistance is desirable.

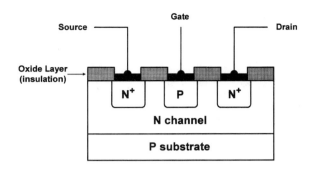

Figure 5.28 N-channel junction gate JFET construction

FET characteristics

As with bipolar transistors, the characteristics of a FET are often presented in the form of a set of graphs relating voltage and current present at the transistor's terminals.

A typical **mutual characteristic** (I_D plotted against V_{GS}) for a small-signal general-purpose N-channel JFET operating in common-source mode (see Chapter 7) is shown in Fig. 5.29. This characteristic shows that the drain current is progressively reduced as the gate–source voltage is made more negative. At a certain value of V_{GS} the drain current falls to zero and the device is said to be **cut-off**.

Figure 5.30 shows a typical **output characteristic** (I_D plotted against V_{GS}) for a small-signal general-purpose N-channel JFET operating in common-source mode (see Chapter 7). This characteristic comprises a family of curves, each relating to a different value of gate–source voltage (V_{GS}). Once again, it is worth taking a little time to get familiar with this characteristic as we shall be using it again in Chapter 7 (you might also like to compare this characteristic with the output characteristic for a transistor operating in common-emitter mode – see Fig. 5.23).

Once again, the characteristic curves have a 'knee' that occurs at low values of V_{DS}. Also, note how the curves become flattened above this value with the drain current (I_D) not changing very greatly for a comparatively large change in drain-source voltage (V_{DS}). These characteristics are, in fact, even flatter than those for a bipolar transistor. Because of their flatness, they are often said to represent a **constant current** characteristic.

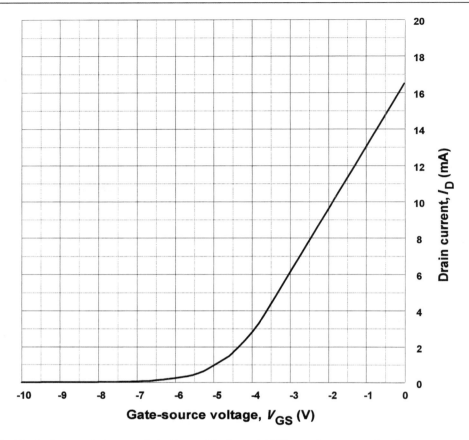

Figure 5.29   Typical mutual characteristic for an N-channel FET transistor operating in common-source mode

FET parameters

The gain offered by a field effect transistor is normally expressed in terms of its **forward transfer conductance** (g_{fs} or Y_{fs}) in **common source** mode. In this mode, the input voltage is applied to the gate and the output current appears in the drain (the source is effectively common to both the input and output circuits).

The common source forward transfer conductance is given by:

$$g_{fs} = I_d/V_{gs}$$

where I_d is the change in drain current resulting from a corresponding change in gate–source voltage (V_{gs}). The units of forward transfer conductance are siemens (S).

Forward transfer conductance (g_{fs}) varies with drain current collector current. For most small-signal devices, g_{fs} is quoted for values of drain current between 1 mA and 10 mA.

Most FET parameters (particularly forward transfer conductance) are liable to wide variation from one device to the next. It is, therefore, important to design circuits on the basis of the minimum value for g_{fs} in order to ensure successful operation with a variety of different devices.

The characteristics of several common N-channel field effect transistors are shown in Table 5.13.

Figure 5.31 shows a test circuit for obtaining the characteristics of an N-channel FET (the arrangement for a P-channel FET is similar but all meters and supplies must be reversed).

Example 5.9

A FET operates with a drain current of 100 mA and a gate–source bias of −1 V. If the device has

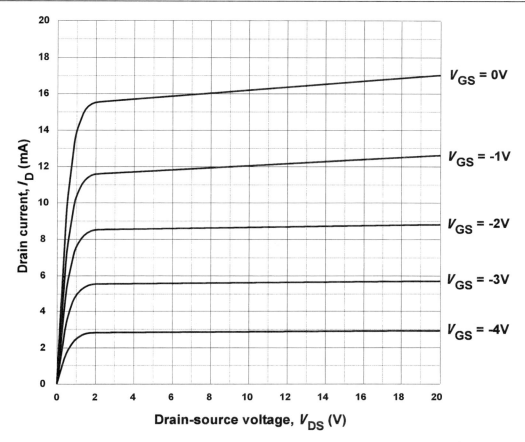

Figure 5.30 Typical family of output characteristic for an N-channel FET transistor operating in common-source mode

Table 5.12 FET parameters

$I_D$max.	– the maximum drain current
V_{DS}max.	– the maximum drain–source voltage
V_{GS}max.	– the maximum gate–source voltage
$P_D$max.	– the maximum drain power dissipation
$t_r$typ.	– the typical output rise-time in response to a perfect rectangular pulse input
$t_f$typ.	– the typical output fall-time in response to a perfect rectangular pulse input
$R_{DS(on)}$max.	– the maximum value of resistance between drain and source when the transistor is in the conducting (on) state

Table 5.13 Characteristics of some common types of junction gate FET

Device	Type	I_D max. (mA)	V_{DS} max. (V)	P_D max. (mW)	g_{fs} min. (mS)	Application
2N3819	N-chan.	10	25	200	4	General purpose
2N5457	N-chan.	10	25	310	1	General purpose
BF244A	N-chan.	100	30	360	3	RF amplifier

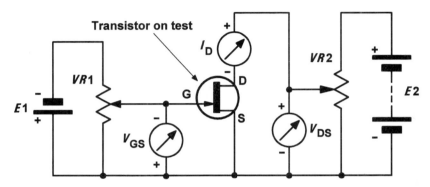

Figure 5.31 N-channel FET test circuit (the arrangement for a P-channel transistor is similar but all meters and supplies must be reversed)

a g_{fs} of 0.25 S, determine the change in drain current if the bias voltage increases to −1.1 V.

Solution

The change in gate–source voltage (V_{gs}) is −0.1 V and the resulting change in drain current can be determined from:

$$I_d = g_{fs} \times V_{gs} = 0.25 \text{ S} \times -0.1 \text{ V} = -0.025 \text{ A}$$
$$= -25 \text{ mA}$$

The new value of drain current will thus be (100 mA − 25 mA) or 75 mA.

Transistor packages

A wide variety of packaging styles are used for transistors. Small-signal transistors tend to have either plastic packages (e.g. TO92) or miniature metal cases (e.g. TO5 or TO18). Medium and high-power devices may also be supplied in plastic cases but these are normally fitted with integral metal heat-sinking tabs (e.g. TO126, TO218 or TO220) in order to conduct heat away from the junction. Some older power transistors are supplied in metal cases (either TO66 or TO3). Several popular transistor case styles are shown in Fig. 5.32.

Integrated circuits

Integrated circuits are complex circuits fabricated on a small slice of silicon. Integrated circuits may contain as few as 10 or more than 100 000 active devices (transistors and diodes). With the exception of a few specialized applications (such as amplification at high power levels) integrated circuits have largely rendered a great deal of conventional discrete circuitry obsolete.

Integrated circuits can be divided into two general classes, **linear** (analogue) and **digital**. Typical examples of linear integrated circuits are operational amplifiers (see Chapter 8) whereas typical examples of digital integrated are logic gates (see Chapter 9). A number of devices bridge the gap between the analogue and digital world. Such devices include analogue to digital converters (ADC), digital to analogue converters (DAC), and timers. For example, the ubiquitous 555 timer contains two

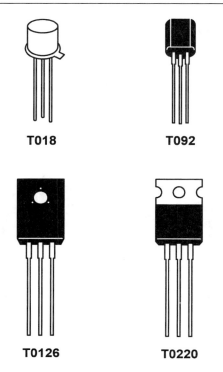

Figure 5.32   Some common transistor packages

operational amplifier stages (configured as voltage comparators) together with a digital bistable stage, a buffer amplifier and an open-collector transistor.

IC packages

As with transistors, a variety of different packages are used for integrated circuits. The most popular form of encapsulation used for integrated circuits is the dual-in-line (DIL) package which may be fabricated from either plastic or ceramic material (with the latter using a glass hermetic sealant). Common DIL packages have 8, 14, 16, 28 and 40 pins on a 0.1 inch matrix.

Flat package (flatpack) construction (featuring both glass–metal and glass–ceramic seals and welded construction) are popular for planar mounting on flat circuit boards. No holes are required to accommodate the leads of such devices which are arranged on a 0.05 inch pitch (i.e. half the pitch used with DIL devices). Single-in-line (SIL) and quad-in-line (QIL) packages are also becoming increasingly popular while TO5, TO72, TO3 and TO220 encapsulations are also found (the latter being commonly used for three-terminal voltage regulators), see Fig. 5.33.

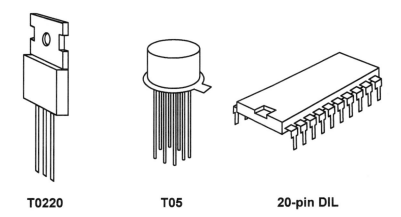

Figure 5.33   Some common integrated circuit packages

Circuit symbols introduced in this chapter

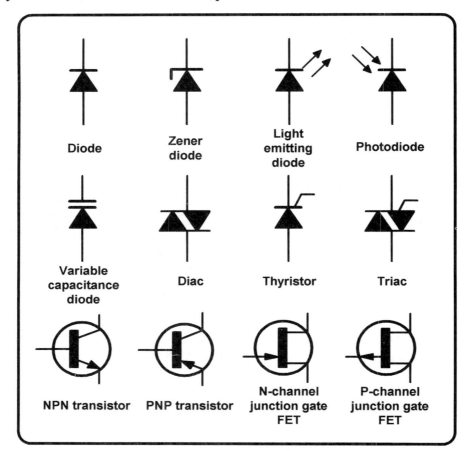

Figure 5.34

Formulae introduced in this chapter

LED series resistor:
(page 93)

$R = (V - V_F)/I$

Bipolar transistor currents:
(page 96)

$I_E = I_B + I_C$

Static current gain for a bipolar transistor:
(page 97)

$h_{FE} = I_C/I_B$

Small signal current gain for a bipolar transistor:
(page 98)

$h_{fe} = I_c/I_b$

Forward transfer conductance for a FET:
(page 102)

$g_{fs} = I_d/V_{gs}$

Problems

5.1   Figure 5.35 shows the characteristics of a diode. What type of material is used in this diode? Give a reason for your answer.

5.2   Use the characteristic shown in Fig. 5.35 to determine the resistance of the diode when (a) $V_F = 0.65$ V and (b) $I_F = 4$ mA.

5.3   The following data refers to a signal diode:

V_F (V): 0.0 0.1   0.2   0.3 0.4 0.5   0.6   0.7
I_F (mA): 0.0 0.05 0.02 1.2 3.6 6.5 10.1 13.8

Plot the characteristic and use it to determine:

Forward current (mA)

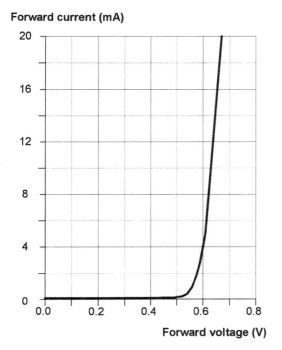

Forward voltage (V)

Figure 5.35

(a) the forward current when $V_F = 350$ mV;
(b) the forward voltage when $I_F = 15$ mA.

5.4   A diode is marked 'BZY88C9V1'. What type of diode is it? What is its rated voltage? State one application for the diode.

5.5   An LED is to be used to indicate the presence of a 5 V d.c. supply. If the LED has a nominal forward voltage of 2 V, and is rated at a current of 12 mA, determine the value of series resistor required.

5.6   Identify each of the following transistors:

(a) AF117   (b) BC184
(c) BD131   (d) BF180.

5.7   A transistor operates with a collector current of 2.5 A and a base current 125 mA. Determine the value of emitter current and static common-emitter current gain.

5.8   A transistor operates with a collector current of 98 mA and an emitter current of 103 mA. Determine the value of base current and the static value of common-emitter current gain.

5.9   A bipolar transistor is to be used in a driver circuit in which a base current of 12 mA is available. If the load requires a current of

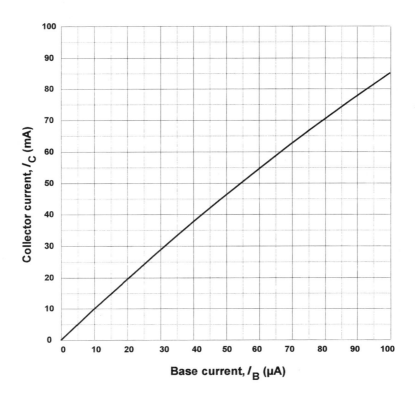

Figure 5.36

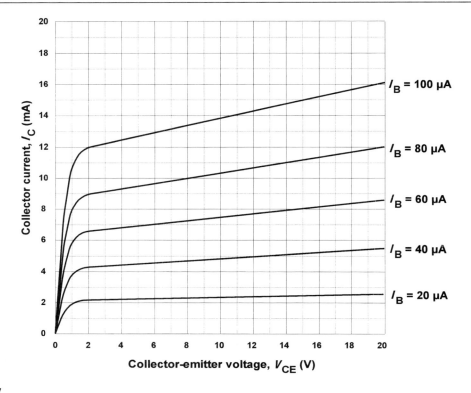

Figure 5.37

200 mA, determine the minimum value of common-emitter current gain required.

5.10   An NPN transistor is to operate with $V_{CE} = 10$ V, $I_C = 50$ mA, and $I_B = 400$ μA. Which of the devices listed in Table 5.11 is most suitable for use in this application.

5.11   A transistor is used in a linear amplifier arrangement. The transistor has small and large signal current gains of 250 and 220, respectively, and bias is arranged so that the static value of collector current is 2 mA. Determine the value of base bias current and the change of output (collector) current that would result from a 5 μA change in input (base) current.

5.12   The transfer characteristic for an NPN transistor is shown in Fig. 5.36. Use this characteristic to determine:

(a) I_C when $I_B = 50$ μA;
(b) h_{FE} when $I_B = 50$ μA;
(c) h_{fe} when $I_C = 75$ mA.

5.13   The output characteristic of an NPN transistor is shown in Fig. 5.37. Use this characteristic to determine:

(a) I_C when $I_B = 100$ μA and $V_{CE} = 4$ V;
(b) V_{CE} when $I_B = 40$ μA and $I_C = 5$ mA;
(c) I_B when $I_C = 7$ mA and $V_{CE} = 6$ V.

5.14   An N-channel FET operates with a drain current of 20 mA and a gate–source bias of −1 V. If the device has a g_{fs} of 8 mS, determine the new drain current if the bias voltage increases to −1.5 V.

5.15   The following results were obtained during an experiment on an N-channel FET:

V_{GS} (V)	I_D (mA)
0	11.4
−2	7.6
−4	3.8
−6	0.2
−8	0

Plot the mutual characteristic for the FET and use it to determine g_{fs} when $I_D = 5$ mA.

(Answers to these problems appear on page 202.)

6

Power supplies

This chapter deals with the unsung hero of most electronic systems, the power supply. Nearly all electronic circuits require a source of well regulated d.c. at voltages of typically between 5 V and 30 V. In some cases, this supply can be derived directly from batteries (e.g. 6 V, 9 V, 12 V) but in many others it is desirable to make use of a standard a.c. mains outlet. This chapter explains how rectifier and smoothing circuits operate and how power supply output voltages can be closely regulated. The chapter concludes with a brief description of some practical power supply circuits.

The block diagram of a d.c. power supply is shown in Fig. 6.1. Since the mains input is at a relatively high voltage, a step-down transformer of appropriate turns ratio is used to convert this to a low voltage. The a.c. output from the transformer secondary is then rectified using conventional silicon rectifier diodes (see Chapter 5) to produce an unsmoothed (sometimes referred to as **pulsating d.c.**) output. This is then smoothed and filtered before being applied to a circuit which will **regulate** (or **stabilize**) the output voltage so that it remains relatively constant in spite of variations in both load current and incoming mains voltage.

Figure 6.2 shows how some of the electronic components that we have already met can be used in the realization of the block diagram in Fig. 6.1. The iron-cored stepdown transformer feeds a rectifier arrangement (often based on a bridge circuit). The output of the rectifier is then applied to a high-value **reservoir** capacitor. This capacitor stores a considerable amount of charge and is being constantly topped-up by the rectifier arrangement. The capacitor also helps to smooth out the voltage pulses produced by the rectifier. Finally, a stabilizing circuit (often based on a **series transistor regulator** and a zener diode **voltage reference**) provides a constant output voltage.

We shall now examine each stage of this arrangement in turn, building up to some complete power supply circuits at the end of the chapter.

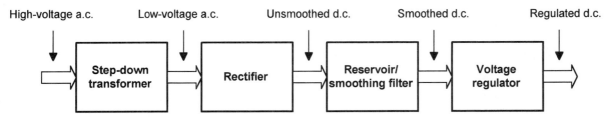

Figure 6.1   Block diagram of a d.c. power supply

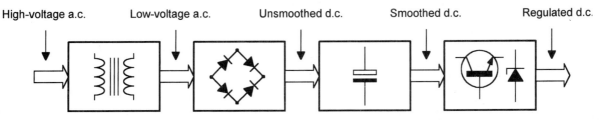

Figure 6.2   Block diagram of a d.c. power supply showing principal components used in each stage

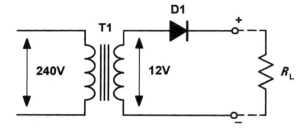

Figure 6.3   Simple half-wave rectifier circuit

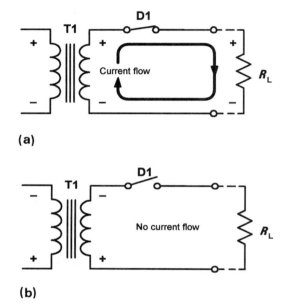

(a)

(b)

Figure 6.4   (a) Half-wave rectifier circuit with D1 conducting (positive-going half-cycles of secondary voltage) (b) half-wave rectifier circuit with D1 not conducting (negative-going half-cycles of secondary voltage)

Rectifiers

Semiconductor diodes (see Chapter 5) are commonly used to convert alternating current (a.c.) to direct current (d.c.), in which case they are referred to as **rectifiers**. The simplest form of rectifier circuit makes use of a single diode and, since it operates on only either positive or negative half-cycles of the supply, it is known as a **half-wave** rectifier.

Figure 6.3 shows a simple half-wave rectifier circuit. Mains voltage (240 V) is applied to the primary of a stepdown transformer (T1). The secondary of T1 steps down the 240 V r.m.s. to 12 V r.m.s. (the turns ratio of T1 will thus be 240/12 or

20:1). Diode D1 will only allow the current to flow in the direction shown (i.e. from cathode to anode). D1 will be forward biased during each positive half-cycle (relative to common) and will effectively behave like a closed switch. When the circuit current tries to flow in the opposite direction, the voltage bias across the diode will be reversed, causing the diode to act like an open switch (see Figs 6.4(a) and 6.4(b), respectively).

The switching action of D1 results in a pulsating output voltage which is developed across the load resistor (R_L). Since the mains supply is at 50 Hz, the pulses of voltage developed across R_L will also be at 50 Hz even if only half the a.c. cycle is present. During the positive half-cycle, the diode will drop the 0.6 V to 0.7 V forward threshold voltage normally associated with silicon diodes. However, during the negative half-cycle the peak a.c. voltage will be dropped across D1 when it is reverse biased. This is an important consideration when selecting a diode for a particular application.

Assuming that the secondary of T1 provides 12 V r.m.s., the peak voltage output from the transformer's secondary winding will be given by:

$$V_{pk} = 1.414 \times V_{r.m.s.} = 1.414 \times 12 \text{ V} = 16.968 \text{ V}$$

The peak voltage applied to D1 will thus be approximately 17 V. The negative half-cycles are blocked by D1 and thus only the positive half-cycles appear across R_L. Note, however, that the actual peak voltage across R_L will be the 17 V positive peak being supplied from the secondary on T1, **minus** the 0.7 V forward threshold voltage dropped by D1. In other words, positive half-cycle pulses having a peak amplitude of 16.3 V will appear across R_L.

Example 6.1

A mains transformer having a turns ratio of 11:1 is connected to a 220 V r.m.s. mains supply. If the secondary output is applied to a half-wave rectifier, determine the peak voltage that will appear across a load.

Solution

The r.m.s. secondary voltage will be given by:

$$V_S = V_P/11 = 220/44 = 5 \text{ V}$$

The peak voltage developed after rectification will be given by:

$$V_{PK} = 1.414 \times 20 \text{ V} = 7.07 \text{ V}$$

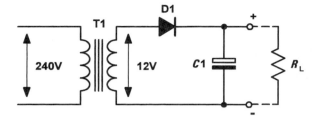

Figure 6.5 Half-wave rectifier with reservoir capacitor

Assuming that the diode is a silicon device with a forward voltage drop of 0.7 V, the actual peak voltage dropped across the load will be:

$$V_L = 7.07 \text{ V} - 0.7 \text{ V} = 6.37 \text{ V}$$

Reservoir and smoothing circuits

Figure 6.5 shows a considerable improvement to the circuit of Fig. 6.3. The capacitor, $C1$, has been added to ensure that the output voltage remains at, or near, the peak voltage even when the diode is not conducting. When the primary voltage is first applied to T1, the first positive half-cycle output from the secondary will charge $C1$ to the peak value seen across R_L. Hence $C1$ charges to 16.3 V at the peak of the positive half-cycle. Because $C1$ and R_L are in parallel, the voltage across R_L will be the same as that across $C1$.

The time required for $C1$ to charge to the maximum (peak) level is determined by the charging circuit time constant (the series resistance multiplied by the capacitance value). In this circuit, the series resistance comprises the secondary winding resistance together with the forward resistance of the diode and the (minimal) resistance of the wiring and connections. Hence $C1$ charges very rapidly as soon as D1 starts to conduct.

The time required for $C1$ to discharge is, in contrast, very much greater. The discharge time constant is determined by the capacitance value and the load resistance, R_L. In practice, R_L is very much larger than the resistance of the secondary circuit and hence $C1$ takes an appreciable time to discharge. During this time, D1 will be reverse biased and will thus be held in its non-conducting state. As a consequence, the only discharge path for $C1$ is through R_L.

$C1$ is referred to as a **reservoir** capacitor. It stores charge during the positive half-cycles of secondary voltage and releases it during the negative half-cycles. The circuit of Fig. 6.5 is thus able to maintain a reasonably constant output voltage across R_L. Even so, $C1$ will discharge by a small amount during the negative half-cycle periods from the transformer secondary. Figure 6.6 shows the secondary voltage waveform together with the voltage developed across R_L with and without $C1$ present. This gives rise to a small variation in the d.c. output voltage (known as **ripple**).

Since ripple is undesirable we must take additional precautions to reduce it. One obvious method of reducing the amplitude of the ripple is that of simply increasing the discharge time constant. This can be achieved either by increasing the value of $C1$ or by increasing the resistance value of R_L. In practice, however, the latter is not really an option

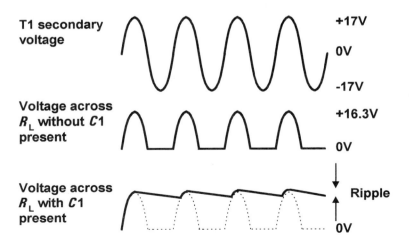

Figure 6.6 Waveforms for half-wave rectifier circuits

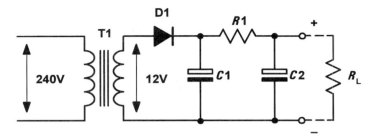

Figure 6.7   Half-wave rectifier with $R–C$ smoothing circuit

because R_L is the effective resistance of the circuit being supplied and we don't usually have the ability to change it! Increasing the value of $C1$ is a more practical alternative and very large capacitor values (often in excess of 4700 µF) are typical.

Figure 6.7 shows a further refinement of the simple power supply circuit. This circuit employs two additional components, $R1$ and $C2$, which act as a filter to remove the ripple. The value of $C2$ is chosen so that the component exhibits a negligible reactance at the ripple frequency (50 Hz for a half-wave rectifier or 100 Hz for a full-wave rectifier – see later). In effect, $R1$ and $C2$ act like a potential divider. The amount of ripple is reduced by an approximate factor equal to:

$$\frac{X_C}{\sqrt{(R1^2 + X_C^2)}}$$

Example 6.2

The $R–C$ smoothing filter in a 50 Hz mains operated half-wave rectifier circuit consists of $R1 = 100$ Ω and $C2 = 1000$ µF. If 1 V of ripple appears at the input of the circuit, determine the amount of ripple appearing at the output.

Solution

First we must determine the reactance of the capacitor, $C2$, at the ripple frequency (50 Hz):

$$X_C = \frac{1}{2\pi fC} = \frac{1}{6.28 \times 50 \times 1000 \times 10^{-6}}$$

$$= \frac{1000}{6.28 \times 50} = \frac{1000}{314}$$

Thus $X_C = 3.18$ Ω.

The amount of ripple at the output of the circuit (i.e. appearing across $C2$) will be given by:

$$V_{ripple} = 1 \text{ V} \times \frac{X_C}{\sqrt{(R1^2 + X_C^2)}}$$

$$= 1 \text{ V} \times \frac{3.18}{\sqrt{(100^2 + 3.18^2)}}$$

$$= 0.032 \text{ V} = 32 \text{ mV}$$

Improved ripple filters

A further improvement can be achieved by using an inductor ($L1$) instead of a resistor ($R1$) in the smoothing circuit. This circuit also offers the advantage that the minimum d.c. voltage is dropped across the inductor (in the circuit of Fig. 6.7, the d.c. output voltage is *reduced* by an amount equal to the voltage drop across $R1$).

Figure 6.8 shows the circuit of a half-wave power supply with an $L–C$ smoothing circuit. At the ripple frequency, $L1$ exhibits a high value of inductive reactance while $C1$ exhibits a low value of capacitive reactance. The combined effect is that of an attenuator which greatly reduces the amplitude of the ripple while having a negligible effect on the direct voltage.

Example 6.3

The $L–C$ smoothing filter in a 50 Hz mains operated half-wave rectifier circuit consists of $L1 = 10$ H and $C2 = 1000$ µF. If 1 V of ripple appears at the input of the circuit, determine the amount of ripple appearing at the output.

Solution

Once again, the reactance of the capacitor, $C2$, is 3.18 Ω (see Example 6.2).

The reactance of $L1$ at 50 Hz can be calculated from:

$$X_L = 2\pi fL = 2 \times 3.14 \times 50 \times 10 = 3140 \text{ } \Omega$$

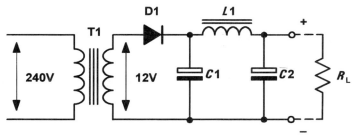

Figure 6.8   Half-wave rectifier with *L–C* smoothing circuit

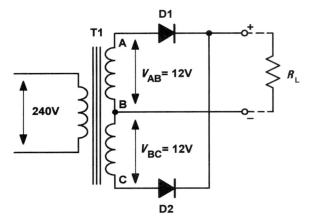

Figure 6.9   Bi-phase rectifier

The amount of ripple at the output of the circuit (i.e. appearing across *C*2) will be approximately given by:

$$V_{\text{ripple}} = 1 \text{ V} \times \frac{X_C}{X_L + X_C}$$

$$= 1 \text{ V} \times \frac{3.18}{3142 + 3.18}$$

$$\approx 0.001 \text{ V or 1 mV}$$

It is worth comparing this value with that obtained from the previous example!

Full-wave rectifiers

The half-wave rectifier circuit is relatively inefficient as conduction takes place only on alternate half-cycles. A better rectifier arrangement would make use of both positive **and** negative half-cycles. These full-wave rectifier circuits offer a considerable improvement over their half-wave counterparts. They are not only more efficient but are significantly less demanding in terms of the reservoir and smoothing components. There are two basic forms of full-wave rectifier; the bi-phase type and the bridge rectifier type.

Bi-phase rectifier circuits

Figure 6.9 shows a simple bi-phase rectifier circuit. Mains voltage (240 V) is applied to the primary of a stepdown transformer (T1) which has two identical secondary windings, each providing 12 V r.m.s. (the turns ratio of T1 will thus be 240/12 or 20:1 for *each* secondary winding).

On positive half-cycles, point A will be positive with respect to point B. Similarly, point B will be positive with respect to point C. In this condition D1 will allow conduction (its anode will be positive with respect to its cathode) while D2 will not allow conduction (its anode will be negative with respect to its cathode). Thus D1 alone conducts on positive half-cycles.

On negative half-cycles, point C will be positive with respect to point B. Similarly, point B will be positive with respect to point A. In this condition D2 will allow conduction (its anode will be positive with respect to its cathode) while D1 will not allow conduction (its anode will be negative with respect to its cathode). Thus D2 alone conducts on negative half-cycles.

Figure 6.10 shows the bi-phase rectifier circuit with the diodes replaced by switches. In Fig. 6.10(a) D1 is shown conducting on a positive half-cycle while in Figure 6.10(b) D2 is shown conducting. The result is that current is routed through the load *in the same direction* on successive half-cycles. Furthermore, this current is derived alternately from the two secondary windings.

As with the half-wave rectifier, the switching action of the two diodes results in a pulsating output voltage being developed across the load resistor (*R*_L). However, unlike the half-wave circuit the

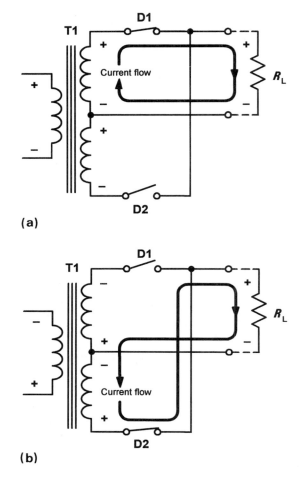

(a)

(b)

Figure 6.10 (a) Bi-phase rectifier with D1 conducting and D2 non-conducting. (b) Bi-phase rectifier with D2 conducting and D1 non-conducting

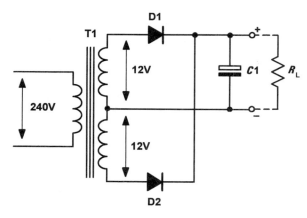

Figure 6.11 Bi-phase rectifier with reservoir capacitor

pulses of voltage developed across R_L will occur at a frequency of 100 Hz (**not** 50 Hz). This doubling of the ripple frequency allows us to use smaller values of reservoir and smoothing capacitor to obtain the same degree of ripple reduction (recall that the reactance of a capacitor is reduced as frequency increases).

As before, the peak voltage produced by each of the secondary windings will be approximately 17 V and the peak voltage across R_L will be 16.3 V (i.e. 17 V less the 0.7 V forward threshold voltage dropped by the diodes).

Figure 6.11 shows how a reservoir capacitor ($C1$) can be added to ensure that the output voltage remains at, or near, the peak voltage even when the diodes are not conducting. This component operates in exactly the same way as for the half-wave circuit, i.e. it charges to approximately 16.3 V at the peak of the positive half-cycle and holds the voltage at this level when the diodes are in their non-conducting states.

The time required for $C1$ to charge to the maximum (peak) level is determined by the charging circuit time constant (the series resistance multiplied by the capacitance value). In this circuit, the series resistance comprises the secondary winding resistance together with the forward resistance of the diode and the (minimal) resistance of the wiring and connections. Hence $C1$ charges very rapidly as soon as either D1 or D2 starts to conduct.

The time required for $C1$ to discharge is, in contrast, very much greater. The discharge time constant is determined by the capacitance value and the load resistance, R_L. In practice, R_L is very much larger than the resistance of the secondary circuit and hence $C1$ takes an appreciable time to discharge. During this time, D1 and D2 will be reverse biased and held in a non-conducting state. As a consequence, the only discharge path for $C1$ is through R_L. Figure 6.12 shows voltage waveforms for the bi-phase rectifier, with and without $C1$ present.

Bridge rectifier circuits

An alternative to the use of the bi-phase circuit is that of using a four-diode bridge rectifier (see Fig. 6.13) in which opposite pairs of diode conduct on alternate half-cycles. This arrangement avoids the need to have two separate secondary windings.

A full-wave bridge rectifier arrangement is shown in Fig. 6.14. Mains voltage (240 V) is applied to

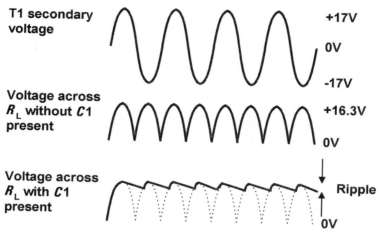

Figure 6.12   Waveforms for the bi-phase rectifier

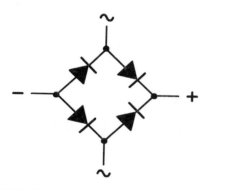

Figure 6.13   Four diodes connected as a bridge rectifier

the primary of a stepdown transformer (T1). The secondary winding provides 12 V r.m.s. (approximately 17 V peak) and has a turns ratio of 20:1, as before. On positive half-cycles, point A will be positive with respect to point B. In this condition D1 and D2 will allow conduction while D3 and D4 will not allow conduction. Conversely, on negative half-cycles, point B will be positive with respect to point A. In this condition D3 and D4 will allow conduction while D1 and D2 will not allow conduction.

Figure 6.15 shows the bridge rectifier circuit with the diodes replaced by switches. In Fig. 6.15(a) D1 and D2 are conducting on a positive half-cycle while in Figure 6.15(b) D3 and D4 are conducting. Once again, the result is that current is routed through the load *in the same direction* on successive half-cycles.

As with the bi-phase rectifier, the switching action of the two diodes results in a pulsating output voltage being developed across the load resistor (R_L). Once again, the peak output voltage is approximately 16.3 V (i.e. 17 V less the 0.7 V forward threshold voltage).

Figure 6.16 shows how a reservoir capacitor ($C1$) can be added to ensure that the output voltage remains at, or near, the peak voltage even when the diodes are not conducting. This component operates

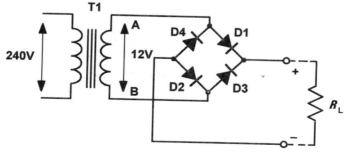

Figure 6.14   Full-wave bridge rectifier circuit

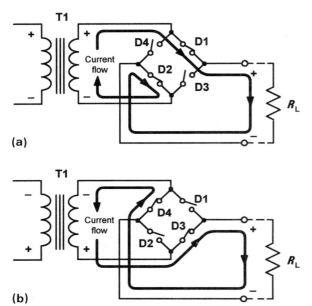

(a)

(b)

Figure 6.15 (a) Bridge rectifier with D1 and D2 conducting, D3 and D4 non-conducting (b) bridge rectifier with D1 and D2 non-conducting, D3 and D4 non-conducting

in exactly the same way as for the bi-phase circuit, i.e. it charges to approximately 16.3 V at the peak of the positive half-cycle and holds the voltage at this level when the diodes are in their non-conducting states. This component operates in exactly the same was as for the bi-phase circuit and the voltage waveforms are identical to those shown in Fig. 6.12.

Finally, R–C and L–C ripple filters can be added to bi-phase and bridge rectifier circuits in exactly the same way as those shown for the half-wave rectifier arrangement (see Figs 6.7 and 6.8).

Voltage regulators

A simple voltage regulator is shown in Fig. 6.17. R_S is included to limit the zener current to a safe value when the load is disconnected. When a load (R_L) is connected, the zener current (I_Z) will fall as current is diverted into the load resistance (it is usual to allow a minimum current of 2 mA to 5 mA in order to ensure that the diode regulates). The output voltage (V_O) will remain at the zener voltage until regulation fails at the point at which the potential divider formed by R_S and R_L produces a lower output voltage that is less than V_Z. The ratio of R_S to R_L is thus important.

At the point at which the circuit just begins to fail to regulate:

$$V_Z = V_{IN} \times \frac{R_L}{R_L + R_S}$$

where V_{IN} is the unregulated input voltage. Thus the *maximum* value for R_S can be calculated from:

$$R_S\text{max} = R_L \times \left(\frac{V_{IN}}{V_Z} - 1 \right)$$

The power dissipated in the zener diode, $P_Z = I_Z \times V_Z$, hence the minimum value for R_S can be determined from the 'off-load' condition when:

$$R_S\text{min} = \frac{V_{IN} - V_Z}{I_Z} = \frac{V_{IN} - V_Z}{(P_Z\text{max}/V_Z)}$$

$$= \frac{(V_{IN} - V_Z) \times V_Z}{P_Z\text{max}}$$

thus

$$R_S\text{min} = \frac{(V_{IN} \times V_Z) - V_Z^2}{P_Z\text{max}}$$

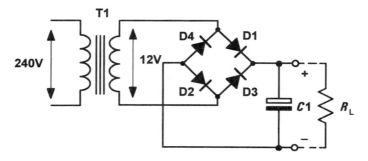

Figure 6.16 Bridge rectifier with reservoir capacitor

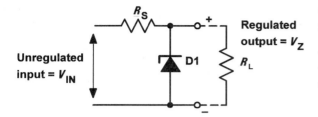

Figure 6.17 Zener diode shunt voltage regulator

where $P_Z$max is the maximum rated power dissipation for the zener diode.

Example 6.4

A 5 V zener diode has a maximum rated power dissipation of 500 mW. If the diode is to be used in a simple regulator circuit to supply a regulated 5 V to a load having a resistance of 400 Ω, determine a suitable value of series resistor for operation in conjunction with a supply of 9 V.

Solution

First we should determine the maximum value for R_S:

$$R_S\text{max} = R_L \times \left(\frac{V_{IN}}{V_Z} - 1 \right)$$

thus

$$R_S\text{max} = 400 \ \Omega \times \left(\frac{9 \text{ V}}{5 \text{ V}} - 1 \right) = 400 \times (1.8 - 1)$$

$$= 320 \ \Omega$$

Now determine the minimum value for R_S:

$$R_S\text{min} = \frac{(V_{IN} \times V_Z) - V_Z^2}{P_Z\text{max}} = \frac{(9 \times 5) - 5^2}{0.5}$$

$$= \frac{45 - 25}{0.5} = 40 \ \Omega$$

Hence a suitable value for R_S would be 150 Ω (roughly mid-way between the two extremes).

Output resistance and voltage regulation

In a perfect power supply, the output voltage would remain constant regardless of the current taken by the load. In practice, however, the output voltage falls as the load current increases. To account for

this fact, we say that the power supply has **internal resistance** (ideally this should be zero).

This internal resistance appears at the output of the supply and is defined as the change in output voltage divided by the corresponding change in output current. Hence:

$$R_O = \frac{\text{change in output voltage}}{\text{change in output current}} = \frac{dV_O}{dI_L}$$

(where dV_O represents a small change in output voltage and dI_L represents a corresponding small change in output current)

The **regulation** of a power supply is given by the relationship:

$$\text{regulation} = \frac{\text{change in output voltage}}{\text{change in line (input) voltage}} \times 100\%$$

Ideally, the value of regulation should be very small. Simple shunt zener diode regulators are capable of producing values of regulation of 5% to 10%. More sophisticated circuits based on discrete components produce values of between 1% and 5% and integrated circuit regulators often provide values of 1% or less.

Example 6.5

The following data were obtained during a test carried out on a d.c. power supply:

(i) Load test

Output voltage (no-load) = 12 V
Output voltage (2 A load current) = 11.5 V

(ii) Regulation test

Output voltage (mains input, 220 V) = 12 V
Output voltage (mains input, 200 V) = 11.9 V

Determine (a) the equivalent output resistance of the power supply and (b) the regulation of the power supply.

Solution

The output resistance can be determined from the load test data:

$$R_O = \frac{\text{change in output voltage}}{\text{change in output current}} = \frac{(12 - 11.5 \text{ V})}{(2 - 0)}$$

$$= \frac{0.5}{2} = 0.25 \ \Omega$$

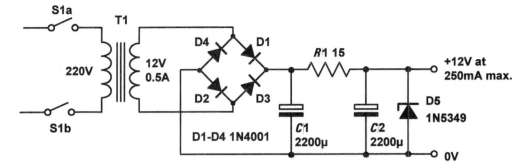

Figure 6.18   Simple d.c. power supply with shunt zener diode regulated output

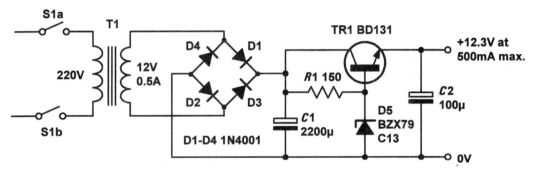

Figure 6.19   Improved regulated d.c. power supply with series-pass transistor

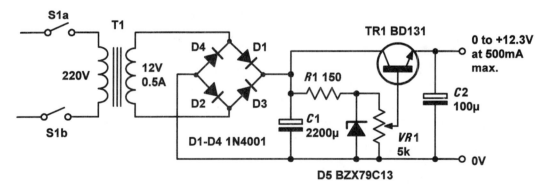

Figure 6.20   Variable d.c. power supply

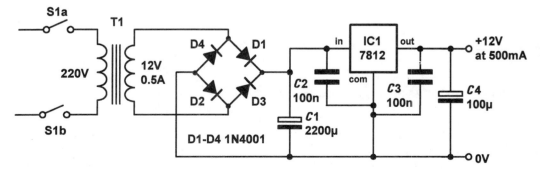

Figure 6.21   Power supply with three-terminal IC voltage regulator

The regulation can be determine from the regulation test data:

$$\text{regulation} = \frac{\text{change in output voltage}}{\text{change in line (input) voltage}} \times 100\%$$

thus

$$\text{regulation} = \frac{(12 - 11.9)}{(220 - 200)} \times 100\%$$

$$= \frac{0.1}{20} \times 100\% = 0.5\%$$

Practical power supply circuits

Figure 6.18 shows a simple power supply circuit capable of delivering an output current of up to 250 mA. The circuit uses a full-wave bridge rectifier arrangement (D1 to D4) and a simple $C–R$ filter. The output voltage is regulated by the shunt-connected 12 V zener diode.

Figure 6.19 shows an improved power supply in which a transistor is used to provide current gain and minimize the power dissipated in the zener diode (TR1 is sometimes referred to as a **series-pass** transistor). The zener diode, D5, is rated at 13 V and the output voltage will be approximately 0.7 V less than this (i.e. 13 V minus the base–emitter voltage drop associated with TR1). Hence the output voltage is about 12.3 V. The circuit is capable of delivering an output current of up to 500 mA (note that TR1 should be fitted with a small heatsink to conduct away any heat produced).

Figure 6.20 shows a variable power supply. The base voltage to the series-pass transistor is derived from a potentiometer connected across the zener diode, D5. Hence the base voltage is variable from 0 V to 13 V. The transistor requires a substantial heatsink (note that TR1's dissipation *increases* as the output voltage is reduced).

Finally, Fig. 6.21 shows a d.c. power supply based on a fixed-voltage three-terminal integrated circuit voltage regulator. These devices are available in standard voltage and current ratings (e.g. 5 V, 12 V, 15 V at 1 A, 2 A and 5 A) and they provide excellent performance in terms of output resistance, ripple rejection and voltage regulation. In addition, such devices usually incorporate over-current protection and can withstand a direct short-circuit placed across their output terminals. This is an essential feature in many practical applications!

Circuit symbols introduced in this chapter

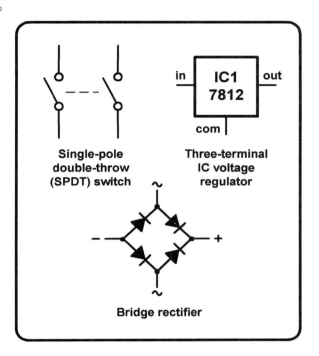

Figure 6.22

Problems

6.1 A half-wave rectifier is fitted with an $R–C$ smoothing filter comprising $R = 200\ \Omega$ and $C = 50\ \mu F$. If 2 V of 400 Hz ripple appear at the input of the circuit, determine the amount of ripple appearing at the output.

6.2 The $L–C$ smoothing filter fitted to a 50 Hz mains operated full-wave rectifier circuit consists of $L = 4$ H and $C = 500\ \mu F$. If 4 V of ripple appear at the input of the circuit, determine the amount of ripple appearing at the output.

6.3 If a 9 V zener diode is to be used in a simple shunt regulator circuit to supply a load having a nominal resistance of 300 Ω, determine the maximum value of series resistor for operation in conjunction with a supply of 15 V.

6.4 The circuit of a d.c. power supply is shown in Fig. 6.23. Determine the voltages that will appear at test points A, B and C.

6.5 In Fig. 6.23, determine the current flowing in

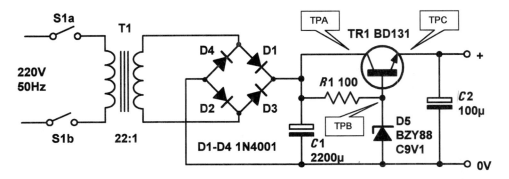

Figure 6.23

$R1$ and the power dissipated in D5 when the circuit is operated without any load connected.

6.6   In Fig. 6.23, determine the effect of each of the following fault conditions:

(a) $R1$ open-circuit;
(b) D5 open-circuit;
(c) D5 short-circuit.

6.7   The following data were obtained during a load test carried out on a d.c. power supply:

Output voltage (no-load) = 8.5 V
Output voltage (800 mA load current) = 8.1 V

Determine the output resistance of the power supply and estimate the output voltage at a load current of 400 mA.

6.8   The following data were obtained during a regulation test on a d.c. power supply:

Output voltage (mains input, 230 V) = 15 V
Output voltage (mains input, 190 V) = 14.6 V

Determine the regulation of the power supply and estimate the output voltage when the input voltage is 245 V.

(Answers to these problems appear on page 202.)

7

Amplifiers

This chapter introduces the basic concepts of amplifiers and amplification. It describes the most common types of amplifier and outlines the basic classes of operation used in both linear and non-linear amplifiers. The chapter also describes methods for predicting the performance of an amplifier based on equivalent circuits and on the use of semiconductor characteristics and load-lines. Once again, we conclude with a selection of practical circuits that can be built and tested.

Types of amplifier

Many different types of amplifier are found in electronic circuits. Before we explain the operation of transistor amplifiers in detail, it is worth describing some of the types of amplifier used in electronic circuits:

AC coupled amplifiers

In a.c. coupled amplifiers, stages are coupled together in such a way that d.c. levels are isolated and only the a.c. components of a signal are transferred from stage to stage.

DC coupled amplifiers

In d.c. (or direct) coupled, stages are coupled together in such a way that stages are not isolated to d.c. poentials. Both a.c. and d.c. signal components are transferred from stage to stage.

Large-signal amplifiers

Large-signal amplifiers are designed to cater for appreciable voltage and/or current levels (typically from 1 V to 100 V or more).

Small-signal amplifiers

Small-signal amplifiers are designed to cater for low level signals (normally less than 1 V and often much smaller).

Audio frequency amplifiers

Audio frequency amplifiers operate in the band of frequencies that is normally associated with audio signals (e.g. 20 Hz to 20 kHz).

Wideband amplifiers

Wideband amplifiers are capable of amplifying a very wide range of frequencies, typically from a few tens of hertz to several megahertz.

Radio frequency amplifiers

Radio frequency amplifiers operate in the band of frequencies that is normally associated with radio signals (e.g. from 100 kHz to over 1 GHz). Note that it is desirable for amplifiers of this type to be **frequency selective** and thus their frequency response may be restricted to a relatively narrow band of frequencies (see Fig. 7.10).

Low-noise amplifiers

Low-noise amplifiers are designed so that they contribute negligible noise (signal disturbance) to the signal being amplified. These amplifiers are usually designed for use with very small signal levels (usually less than 10 mV or so).

Gain

One of the most important parameters of an amplifier is the amount of amplification or **gain** that it provides. Gain is simply the ratio of output voltage to input voltage, output current to input current, or output power to input power (see Fig. 7.1). These three ratios give, respectively, the voltage gain, current gain and power gain. Thus:

voltage gain, $A_V = \dfrac{V_{out}}{V_{in}}$

current gain, $A_i = \dfrac{I_{out}}{I_{in}}$

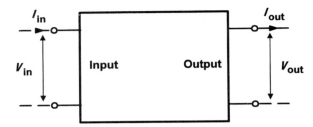

Figure 7.1 Block diagram for an amplifier showing input and output voltages and currents

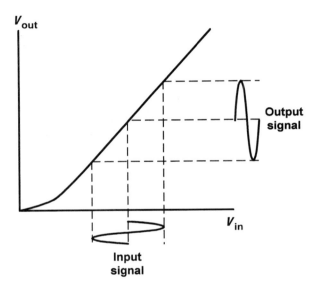

Figure 7.2 Class A (linear) operation

and

power gain, $A_p = \dfrac{P_{out}}{P_{in}}$

Note that, since power is the product of current and voltage ($P = IV$), we can infer that:

$$A_p = \frac{P_{out}}{P_{in}} = \frac{I_{out} \times V_{out}}{I_{in} \times V_{in}} = \frac{I_{out}}{I_{in}} \times \frac{V_{out}}{V_{in}} = A_i \times A_v$$

Hence,

$A_p = A_i \times A_v$

Example 7.1

An amplifier produces an output voltage of 2 V for an input of 50 mV. If the input and output currents in this condition are, respectively, 4 mA and 200 mA, determine:

(a) the voltage gain;
(b) the current gain;
(c) the power gain.

Solution

(a) The voltage gain is calculated from:

$$A_v = \frac{V_{out}}{V_{in}} = \frac{2\ \text{V}}{50\ \text{mV}} = 40$$

(b) The current gain is calculated from:

$$A_i = \frac{I_{out}}{I_{in}} = \frac{200\ \text{mA}}{4\ \text{mA}} = 50$$

(c) The power gain is calculated from:

$$A_p = \frac{P_{out}}{P_{in}} = \frac{V_{out} \times I_{out}}{V_{in} \times I_{in}}$$

$$= \frac{2\ \text{V} \times 200\ \text{mA}}{50\ \text{mV} \times 4\ \text{mA}} = \frac{0.4\ \text{W}}{200\ \mu\text{W}}$$

thus

$A_p = 2000$

(Note that $A_p = A_v \times A_i = 50 \times 40 = 2000$.)

Class of operation

A requirement of most amplifiers is that the output signal should be a faithful copy of the input signal, albeit somewhat larger in amplitude. Other types of amplifier are 'non-linear', in which case their input and output waveforms will not necessarily be similar. In practice, the degree of linearity provided by an amplifier can be affected by a number of factors including the amount of bias applied (see page 123) and the amplitude of the input signal. It is also worth noting that a linear amplifier will become non-linear when the applied input signal exceeds a threshold value. Beyond this value the amplifier is said to be 'over-driven' and the output will become increasingly distorted if the input signal is further increased.

Amplifiers are usually designed to be operated with a particular value of bias supplied to the active devices (i.e. transistors). For linear operation, the active device(s) must be operated in the linear part of their transfer characteristic (V_{out} plotted against V_{in}). In Figure 7.2 the input and output signals for an amplifier are operating in linear mode. This form of operation is known as **Class A** and the bias

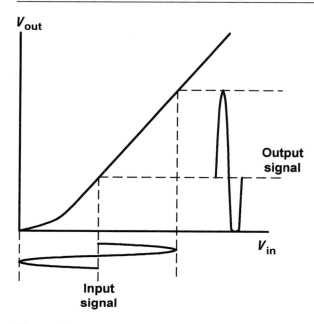

Figure 7.3 Effect of reducing bias and increasing input signal amplitude (the output waveform is no longer a faithful reproduction of the input)

point is adjusted to the mid-point of the linear part of the transfer characteristic. Furthermore, current will flow in the active devices used in a Class A amplifier during a complete cycle of the signal waveform. At no time does the current fall to zero.

Figure 7.3 shows the effect of moving the bias point down the transfer characteristic and, at the same time, increasing the amplitude of the input signal. From this, you should notice that the extreme negative portion of the output signal has become distorted. This effect arises from the non-linearity of the transfer characteristic that occurs near the origin (i.e. the zero point). Despite the obvious non-linearity in the output waveform, the active device(s) will conduct current during a complete cycle of the signal waveform.

Now consider the case of reducing the bias even further while further increasing the amplitude of the input signal (see Fig. 7.4). Here the bias point has been set at the **projected cut-off** point. The negative-going portion of the output signal becomes cut off (or **clipped**) and the active device(s) will cease to conduct for this part of the cycle. This mode of operation is known as **Class AB**.

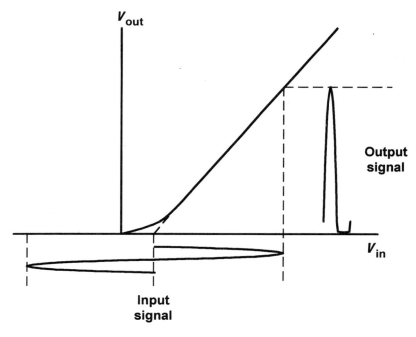

Figure 7.4 Class AB operation (bias set at projected cut-off)

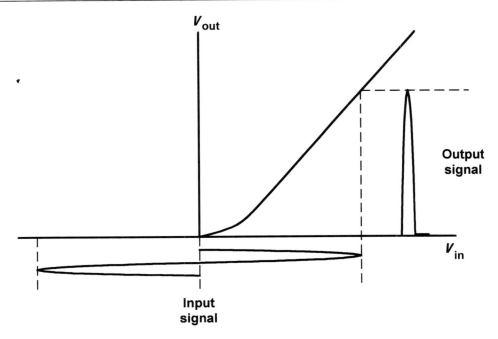

Figure 7.5   Class B operation (no bias applied)

Now let's consider what will happen if no bias at all is applied to the amplifier (see Fig. 7.5). The output signal will only comprise a series of positive-going half-cycles and the active device(s) will only be conducting during half-cycles of the waveform (i.e. they will only be operating 50% of the time). This mode of operation is known as **Class B** and is commonly used in **push–pull** power amplifiers where the two active devices in the output stage operate on alternate half-cycles of the waveform.

Finally, there is one more class of operation to consider. The input and output waveforms for **Class C** operation are shown in Fig. 7.6. Here the bias point is set at beyond the cut-off (zero) point and a very large input signal is applied. The output waveform will then comprise a series of quite sharp positive-going pulses. These pulses of current or voltage can be applied to a tuned circuit load in order to recreate a sinusoidal signal. In effect, the pulses will excite the tuned circuit and its inherent **flywheel action** will produce a sinusoidal output waveform. This mode of operation is only used in RF power amplifiers which must operate at high levels of efficiency.

Table 7.1 summarizes the classes of operation used in amplifiers.

Input and output resistance

Input resistance is the ratio of input voltage to input current and it is expressed in ohms. The input of an amplifier is normally purely resistive (i.e. any reactive component is negligible) in the middle of its working frequency range (i.e. the **mid-band**). In some cases, the reactance of the input may become appreciable (e.g. if a large value of stray capacitance appears in parallel with the input resistance). In such cases we would refer to **input impedance** rather than input resistance.

Output resistance is the ratio of open-circuit output voltage to short-circuit output current and is measured in ohms. Note that this resistance is internal to the amplifier and should not be confused with the resistance of a load connected externally. As with input resistance, the output of an amplifier is normally purely resistive and we can safely ignore any reactive component. If this is not the case, we would once again refer to **output impedance** rather than output resistance.

Figure 7.7 shows how the input and output resistances are 'seen' looking into the input and output terminals, respectively. We shall be returning to this **equivalent circuit** later in this chapter.

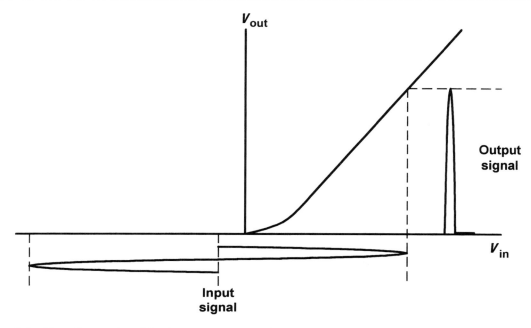

Figure 7.6   Class C operation (bias is set beyond cut-off)

Table 7.1   Classes of operation

Class of operation	Bias point	Conduction angle (typical)	Efficiency (typical)	Application
A	Mid-point	360°	5% to 20%	Linear audio amplifiers
AB	Projected cut-off	210°	20% to 40%	Push–pull audio amplifiers
B	At cut-off	180°	40% to 70%	Push–pull audio amplifiers
C	Beyond cut-off	120°	70% to 90%	Radio frequency power amplifiers

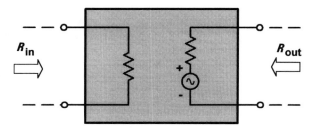

Figure 7.7   Input and output resistances 'seen' looking into the input and output terminals, respectively

Frequency response

The frequency response of an amplifier is usually specified in terms of the upper and lower **cut-off frequencies** of the amplifier. These frequencies are those at which the output power has dropped to 50% (otherwise known as the **–3 dB points**) or where the voltage gain has dropped to 70.7% of its mid-band value. Figures 7.8 and 7.9, respectively, show how the bandwidth can be expressed in terms of power or voltage. In either case, the cut-off frequencies (f_1 and f_2) and bandwidth are identical.

The frequency response characteristics for various types of amplifier are shown in Fig. 7.10. Note

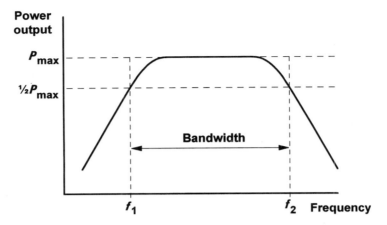

Figure 7.8   Frequency response and bandwidth (output power plotted against frequency)

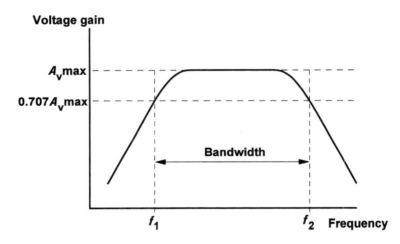

Figure 7.9   Frequency response and bandwidth (output voltage plotted against frequency)

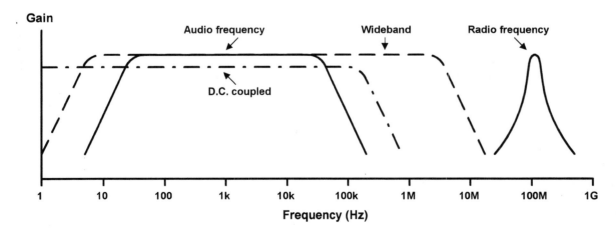

Figure 7.10   Typical frequency response graphs for various types of amplifier

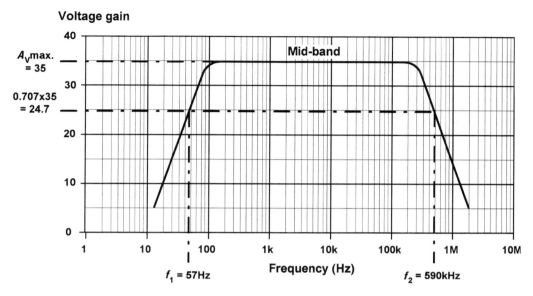

Figure 7.11

that, for response curves of this type, frequency is almost invariably plotted on a logarithmic scale.

Example 7.2

Determine the mid-band voltage gain and upper and lower cut-off frequencies for the amplifier whose frequency response is shown in Fig. 7.11.

Solution

The mid-band voltage gain corresponds with the flat part of the frequency response characteristic. At the point the voltage gain reaches a maximum of 35 (see Fig. 7.11).

The voltage gain at the cut-off frequencies can be calculated from:

$$A_V \text{ cut-off} = 0.707 \times A_V \text{ max}$$
$$= 0.707 \times 35 = 24.7$$

This value of gain intercepts the frequency response at $f_1 = 57$ Hz and $f_2 = 590$ kHz (see Fig. 7.11).

Bandwidth

The bandwidth of an amplifier is usually taken as the difference between the upper and lower cut-off frequencies (i.e. $f_2 - f_1$ in Figs 7.9 and 7.10). The bandwidth of an amplifier must be sufficient to accommodate the range of frequencies present within the signals that it is to be presented with. Many signals contain **harmonic** components (i.e. signals at 2*f*, 3*f*, 4*f*, etc. where *f* is the frequency of the **fundamental** signal). To perfectly reproduce a square wave, for example, requires an amplifier with an infinite bandwidth (note that a square wave comprises an infinite series of harmonics). Clearly it is not possible to perfectly reproduce such a wave but it does explain why it can be desirable for an amplifier's bandwidth to greatly exceed the highest signal frequency that it is required to handle!

Phase shift

Phase shift is the phase angle between the input and output voltages measured in degrees. The measurement is usually carried out in the mid-band where, for most amplifiers, the phase shift remains relatively constant. Note also that conventional single-stage transistor amplifiers usually provide phase shifts of either 180° or 360° (i.e. 0°).

Negative feedback

Many practical amplifiers use negative feedback in order to precisely control the gain, reduce distortion and improve bandwidth. The gain can be reduced to a manageable value by feeding back a small proportion of the output. The amount of feedback

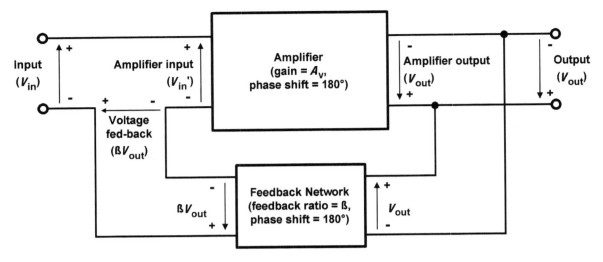

Figure 7.12   Amplifier with negative feedback applied

determines the overall (or **closed-loop**) gain. Because this form of feedback has the effect of reducing the overall gain of the circuit, this form of feedback is known as **negative feedback**. An alternative form of feedback, where the output is fed back in such a way as to reinforce the input (rather than to subtract from it) is known as **positive feedback**. This form of feedback is used in oscillator circuits (see Chapter 9).

Figure 7.12 shows the block diagram of an amplifier stage with negative feedback applied. In this circuit, the proportion of the output voltage fed back to the input is given by β and the overall voltage gain will be given by:

$$\text{overall gain} = \frac{V_{\text{out}}}{V_{\text{in}}}$$

Now $V'_{\text{in}} = V_{\text{in}} - \beta V_{\text{out}}$ (by applying Kirchhoff's voltage law) (note that the amplifier's input voltage has been **reduced** by applying negative feedback) thus

$$V_{\text{in}} = V'_{\text{in}} + \beta V_{\text{out}}$$

and

$$V_{\text{out}} = A_V \times V'_{\text{in}} \;(A_V \text{ is the } \textbf{internal gain} \text{ of the amplifier})$$

Hence,

$$\text{overall gain} = \frac{A_V \times V'_{\text{in}}}{V'_{\text{in}} + \beta V_{\text{out}}} = \frac{A_V \times V'_{\text{in}}}{V'_{\text{in}} + \beta(A_V \times V'_{\text{in}})}$$

Thus

$$\text{overall gain} = \frac{A_V}{1 + \beta A_V}$$

Hence, the overall gain with negative feedback applied will be less than the gain without feedback. Furthermore, if A_V is very large (as is the case with an operational amplifier – see Chapter 8) the overall gain with negative feedback applied will be given by:

$$\text{overall gain (when } A_V \text{ is very large)} = 1/\beta$$

Note, also, that the **loop gain** of a feedback amplifier is defined as the product of β and A_V.

Example 7.3

An amplifier with negative feedback applied has an open-loop voltage gain of 50 and one-tenth of its output is fed back to the input (i.e. β = 0.1). Determine the overall voltage gain with negative feedback applied.

Solution

With negative feedback applied the overall voltage gain will be given by:

$$\frac{A_V}{1 + \beta A_V} = \frac{50}{1 + (0.1 \times 50)} = \frac{50}{1 + 5} = \frac{50}{6} = 8.33$$

Example 7.4

If, in Example 7.3, the amplifier's open-loop voltage gain increases by 20%, determine the percentage increase in overall voltage gain.

Solution

The new value of open-loop gain will be given by:

$$A_V = A_V + 0.2A_V = 1.2 \times 50 = 60$$

The overall voltage gain with negative feedback will then be:

$$\frac{A_V}{1 + \beta A_V} = \frac{60}{1 + (0.1 \times 60)} = \frac{60}{1 + 6} = \frac{50}{6} = 8.57$$

The increase in overall voltage gain, expressed as a percentage, will thus be:

$$\frac{8.57 - 8.33}{8.33} \times 100\% = 2.88\%$$

(Note that this example illustrates one of the important benefits of negative feedback in stabilizing the overall gain of an amplifier stage.)

Example 7.5

An integrated circuit that produces an open-loop gain of 100 is to be used as the basis of an amplifier stage having a precise voltage gain of 20. Determine the amount of feedback required.

Solution

Rearranging the formula, $A_V/(1 + \beta A_V)$, to make β the subject gives:

$$\beta = \frac{1}{A_V'} - \frac{1}{A_V}$$

where A_V' is the overall voltage gain with feedback applied, and A_V is the open-loop voltage gain.

Transistor amplifiers

Regardless of what type of transistor is employed, three basic circuit configurations are used. These three circuit configurations depend upon which one of the three transistor connections is made common to both the input and the output. In the case of bipolar transistors, the configurations are known as **common emitter**, **common collector** (or **emitter follower**) and **common base**. Where field effect transistors are used, the corresponding configurations are **common source**, **common drain** (or **source follower**) and **common gate**.

The three basic circuit configurations (Figs 7.13 to 7.18) exhibit quite different performance charac-

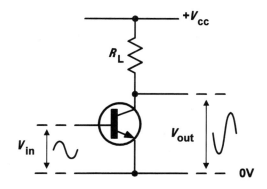

Figure 7.13 Common-emitter configuration

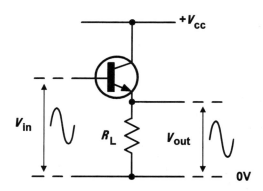

Figure 7.14 Common-collector (emitter follower) configuration

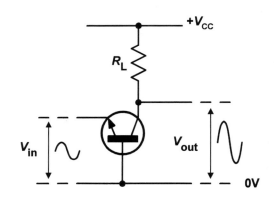

Figure 7.15 Common-base configuration

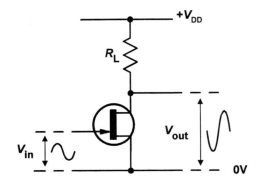

Figure 7.16   Common-source configuration

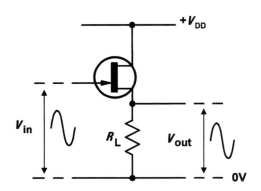

Figure 7.17   Common-drain (source follower) configuration

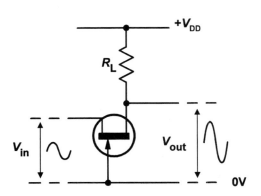

Figure 7.18   Common-gate configuration

teristics, as shown in Tables 7.2 and 7.3 (typical values are given in brackets).

Equivalent circuits

One method of determining the behaviour of an amplifier stage is to make use of an equivalent circuit. Figure 7.19 shows the basic equivalent circuit of an amplifier. The output circuit is reduced to its Thevenin equivalent (see Chapter 3) comprising a voltage generator $(A_V \times V_{in})$ and a series resistance (R_{out}). This simple model allows us to forget the complex circuitry that might exist within the amplifier box!

In practice, we use a slightly more complex equivalent circuit model in the analysis of a transistor amplifier. The most frequently used equivalent circuit is that which is based on **hybrid parameters** (or **h-parameters**). In this form of analysis, a transistor is replaced by four components; h_i, h_r, h_f and h_o (see Table 7.4).

In order to indicate which one of the operating modes is used we add a further subscript letter to each h-parameter; e for common emitter, b for common base and c for common collector (see Table 7.5). However, to keep things simple we shall only consider common-emitter operation.

Common-emitter input resistance (h_{ie})

The input resistance of a transistor is the resistance that is effectively 'seen' between its input terminals. As such, it is the ratio of the voltage between the input terminals to the current flowing into the input. In the case of a transistor operating in common-emitter mode, the input voltage is the voltage developed between the base and emitter, V_{be}, while the input current is the current supplied to the base, I_b.

Figure 7.20 shows the current and voltage at the input of a common-emitter amplifier stage while Fig. 7.21 shows how the input resistance, R_{in}, appears between the base and emitter. Note that R_{in} is not a discrete component – it is *inside* the transistor.

From the foregoing we can deduce that:

$$R_{in} = \frac{V_{be}}{I_b}$$

(note that this is the same as h_{ie}).

Table 7.2 Bipolar transistor amplifier circuit configurations

Parameter	Mode of operation		
	Common emitter (Fig. 7.13)	Common collector (Fig. 7.14)	Common base (Fig. 7.15)
Voltage gain	medium/high (40)	unity (1)	high (200)
Current gain	high (200)	high (200)	unity (1)
Power gain	very high (8000)	high (200)	high (200)
Input resistance	medium (2.5 kΩ)	high (100 kΩ)	low (200 Ω)
Output resistance	medium/high (20 kΩ)	low (100 Ω)	high (100 kΩ)
Phase shift	180°	0°	0°
Typical applications	General purpose, AF and RF amplifiers	Impedance matching, input and output stages	RF and VHF amplifiers

Table 7.3 Field effect transistor amplifier circuit configurations

Parameter	Mode of operation		
	Common source (Fig. 7.16)	Common drain (Fig. 7.17)	Common gate (Fig. 7.18)
Voltage gain	medium (40)	unity (1)	high (250)
Current gain	very high (200 000)	very high (200 000)	unity (1)
Power gain	very high (8 000 000)	very high (200 000)	high (250)
Input resistance	very high (1 MΩ)	very high (1 MΩ)	low (500 Ω)
Output resistance	medium/high (50 kΩ)	low (200 Ω)	high (150 kΩ)
Phase shift	180°	0°	0°
Typical applications	General purpose, AF and RF amplifiers	Impedance matching stages	RF and VHF amplifiers

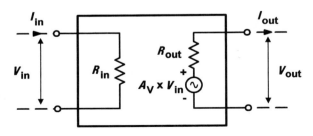

Figure 7.19 Simple equivalent circuit for an amplifier

Table 7.4 Hyrid parameters

h_i	input resistance, $\dfrac{\mathrm{d}v_i}{\mathrm{d}i_i}$
h_r	reverse voltage transfer ratio, $\dfrac{\mathrm{d}v_i}{\mathrm{d}v_o}$
h_f	forward current transfer ratio, $\dfrac{\mathrm{d}i_o}{\mathrm{d}i_i}$
h_o	output conductance, $\dfrac{\mathrm{d}i_o}{\mathrm{d}v_o}$

Table 7.5   *h*-parameters for a transistor operating in common-emitter mode

h_{ie}	input resistance,	$\dfrac{dv_{be}}{di_b}$
h_{re}	reverse voltage transfer ratio,	$\dfrac{dv_{be}}{dv_{ce}}$
h_{fe}	forward current transfer ratio,	$\dfrac{di_c}{di_b}$
h_{oe}	output conductance,	$\dfrac{di_c}{dv_{ce}}$

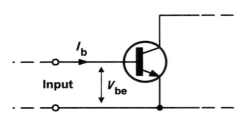

Figure 7.20   Voltage and current at the input of a common-emitter amplifier

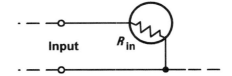

Figure 7.21   Input resistance of a common-emitter amplifier stage

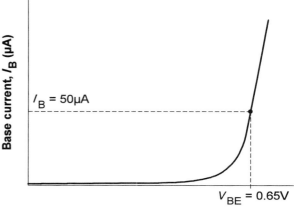

Figure 7.22   Using the input characteristic to determine the large-signal (static) input resistance of a transistor connected in common-emitter mode

The transistor's input characteristic can be used to predict the input resistance of a transistor amplifier stage. Since the input characteristic is non-linear (recall that very little happens until the base–emitter voltage exceeds 0.6 V), the value of input resistance will be very much dependent on the exact point on the graph at which the transistor is being operated. Furthermore, we might expect quite different values of resistance according to whether we are dealing with larger d.c. values or smaller incremental changes (a.c. values). Since this can be a rather difficult concept, it is worth expanding on it.

Figure 7.22 shows a typical input characteristic in which the transistor is operated with a base current, I_B, of 50 µA. This current produces a base–emitter voltage, V_{BE}, of 0.65 V. The input resistance corresponding to these steady (d.c.) values will be given by:

$$R_{in} = \frac{V_{BE}}{I_B} = \frac{0.65\ V}{50\ \mu A} = 13\ k\Omega$$

Now, suppose that we apply a steady bias current of, say, 70 µA and superimpose on this a signal that varies above and below this value, swinging through a total change of 100 µA (i.e. from 20 µA

to 120 µA). Figure 7.23 shows that this produces a base–emitter voltage change of 0.05 V.

The input resistance seen by this small-signal input current is given by:

$$R_{in} = \frac{\text{change in } V_{BE}}{\text{change in } I_B} = \frac{dV_{be}}{dI_b} = \frac{0.05\ V}{100\ \mu A} = 500\ \Omega$$

In other words,

$$h_{ie} = 500\ \Omega\ (\text{since } h_{ie} = \frac{dV_{be}}{dI_b})$$

It is worth comparing this value with the steady (d.c.) value. The appreciable difference is entirely attributable to the shape of the input characteristic!

Common-emitter current gain (h_{fe})

The current gain produced by a transistor is the ratio of output current to input current. In the case of a transistor operating in common-emitter mode,

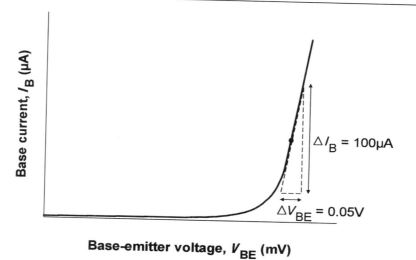

Figure 7.23   Using the input characteristic to determine the small-signal input resistance of a transistor connected in common-emitter mode

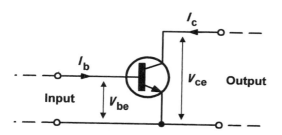

Figure 7.24   Input and output currents and voltages in a common-emitter amplifier stage

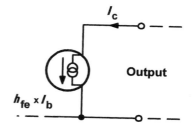

Figure 7.25   Equivalent output current source in a common-emitter amplifier stage

the input current is the base current, I_B, while the output current is the collector current, I_C.

Figure 7.24 shows the input and output currents and voltages for a common-emitter amplifier stage. The magnitude of the current produced at the output of the transistor is equal to the current gain, A_i, multiplied by the applied base current, I_b. Since the output current is the current flowing in the collector, I_c, we can deduce that:

$$I_c = A_i \times I_b$$

where $A_i = h_{fe}$ (the common-emitter current gain).

Figure 7.25 shows how this current source appears between the collector and emitter. Once again, the current source is not a discrete component – it appears *inside* the transistor.

The transistor's transfer characteristic can be used to predict the current gain of a transistor amplifier

stage. Since the transfer characteristic is linear, the current gain remains reasonably constant over a range of collector current.

Figure 7.26 shows a typical transfer characteristic in which the transistor is operated with a base current, I_B, of 240 µA. This current produces a collector current, I_C, of 12 mA. The current gain corresponding to these steady (d.c.) values will be given by:

$$A_i = \frac{I_C}{I_B} = \frac{2.5 \text{ mA}}{50 \text{ µA}} = 50$$

(note that this is the same as h_{FE}, the large-signal or steady-state value of common-emitter current gain).

Now, suppose that we apply a steady bias current of, say, 240 µA and superimpose on this a signal that varies above and below this value, swinging through a total change of 220 µA (i.e. from

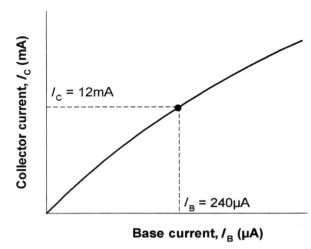

Figure 7.26 Using the transfer characteristic to determine the large-signal (static) current gain of a transistor connected in common-emitter mode

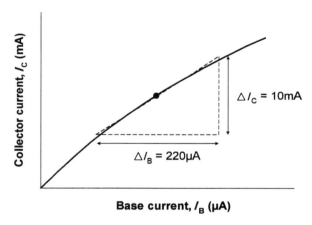

Figure 7.27 Using the transfer characteristic to determine the small-signal current gain of a transistor connected in common-emitter mode

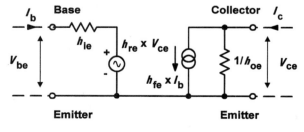

Figure 7.28 *h*-parameter equivalent circuit for a transistor amplifier

Table 7.6   *h*-parameters for a BFY50 transistor

h_{ie} (Ω)	h_{re}	h_{fe}	h_{oe} (μS)
250	0.85×10^{-4}	80	35

(Measured at $I_C = 1$ mA, $V_{CE} = 5$ V.)

120 μA to 360 μA). Figure 7.27 shows that this produces a collector current swing of 10 mA.

The small-signal a.c. current gain is given by:

$$A_i = \frac{\text{change in } I_C}{\text{change in } I_B} = \frac{dI_c}{dI_b} = \frac{10 \text{ mA}}{220 \text{ μA}} = 45.45$$

Once again, it is worth comparing this value with the steady-state value (h_{FE}). Since the transfer characteristic is reasonably linear, the values are quite close (45.45 compared with 50). If the transfer characteristic was perfectly linear, then h_{fe} would be equal to h_{FE}.

h-parameter equivalent circuit for common-emitter operation

The complete *h*-parameter equivalent circuit for a transistor operating in common-emitter mode is shown in Fig. 7.28. We have already shown how the two most important parameters, h_{ie} and h_{fe}, can be found from the transistor's characteristics curves. The remaining parameters, h_{re} and h_{oe}, can, in many applications, be ignored. A typical set of *h*-parameters for a BFY50 transistor is shown in Table 7.6. Note how small h_{re} and h_{oe} are.

Example 7.6

A BFY50 transistor is used in a common-emitter amplifier stage with $R_L = 10$ kΩ and $I_C = 1$ mA. Determine the output voltage produced by an input signal of 10 mV. (You may ignore the effect of h_{re} and any bias components that may be present externally.)

Solution

The equivalent circuit (with h_{re} replaced by a short-circuit) is shown in Fig. 7.29. The load effectively appears between the collector and emitter while the input signal appears between base and emitter.

First we need to find the value of input current, I_b, from:

$$I_b = \frac{V_{in}}{h_{ie}} = \frac{10 \text{ mV}}{250 \text{ } \Omega} = 40 \text{ μA}$$

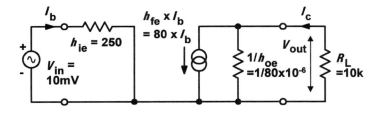

Figure 7.29

Next we find the value of current generated, I_f, from:

$$I_f = h_{fe} \times I_b = 80 \times 40\ \mu A = 320\ \mu A$$

This value of current is shared between the internal resistance between collector and emitter (i.e. $1/h_{oe}$) and the external load, R_L.

To determine the value of collector current, we can apply the current divider theorem (Chapter 3):

$$I_c = I_f \times \frac{1/h_{oe}}{1/h_{oe} + R_L}$$

$$= 320\ \mu A \times \frac{1/(80 \times 10^{-6})}{1/(80 \times 10^{-6}) + 10\ k\Omega}$$

thus

$$I_c = 320\ \mu A \times \frac{12.5\ k\Omega}{12.5\ k\Omega + 10\ k\Omega}$$

$$= 320\ \mu A \times 0.555 = 177.6\ \mu A$$

Finally, we can determine the output voltage from:

$$V_{out} = I_c \times R_L = 177.6\ \mu A \times 10\ k\Omega = 1.776\ V$$

Voltage gain

We can use hybrid parameters to determine the voltage gain of a transistor stage. We have already shown how the voltage gain of an amplifier stage is given by:

$$A_v = \frac{V_{out}}{V_{in}}$$

In the case of a common-emitter amplifier stage, $V_{out} = V_{ce}$ and $V_{in} = V_{be}$. If we assume that h_{oe} and h_{re} are negligible, then:

$$V_{out} = V_{ce} = I_c \times R_L = I_f \times R_L = h_{fe} \times I_b \times R_L$$

and

$$V_{in} = V_{be} = I_b \times R_{in} = I_b \times h_{ie}$$

Thus

$$A_v = \frac{V_{out}}{V_{in}} = \frac{h_{fe} \times I_b \times R_L}{I_b \times h_{ie}} = \frac{h_{fe} \times R_L}{h_{ie}}$$

Example 7.7

A transistor has $h_{fe} = 150$ and $h_{ie} = 1.5\ k\Omega$. Assuming that h_{re} and h_{oe} are both negligible, determine the value of load resistance required to produce a voltage gain of 200.

Solution

Since

$$A_v = \frac{h_{fe} \times R_L}{h_{ie}}, \quad R_L = \frac{A_v \times h_{ie}}{h_{fe}}$$

To produce a gain of 200,

$$R_L = \frac{200 \times 1.5\ k\Omega}{150} = 2\ k\Omega$$

Bias

We stated earlier that the optimum value of bias for a Class A (linear) amplifier is that value which ensures that the active devices are operated at the mid-point of their transfer characteristics. In practice, this means that a static value of collector current will flow even when there is no signal present. Furthermore, the collector current will flow throughout the complete cycle of an input signal (i.e. conduction will take place over an angle of 360°). At no stage will the transistor be **saturated** nor should it be **cut-off**.

In order to ensure that a static value of collector current flows in a transistor, a small current must be applied to the base of the transistor. This current can be derived from the same voltage rail that supplies the collector circuit (via the load). Figure 7.30 shows a simple Class-A common-emitter

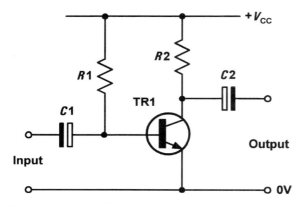

Figure 7.30 Basic Class-A common-emitter amplifier stage

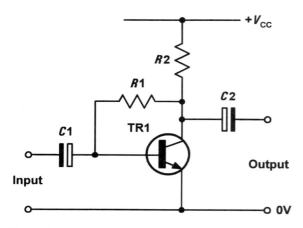

Figure 7.31 Improvement on the circuit shown in Fig. 7.30 (using negative feedback to bias the transistor)

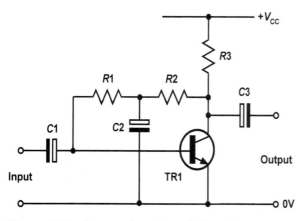

Figure 7.32 Improved version of Fig. 7.31

amplifier circuit in which the base bias resistor, $R1$, and collector load resistor, $R2$, are connected to a common positive supply rail.

The signal is applied to the base terminal of the transistor via a coupling capacitor, $C1$. This capacitor removes the d.c. component of any signal applied to the input terminals and ensures that the base bias current delivered by $R1$ is unaffected by any device connected to the input. $C2$ couples the signal out of the stage and also prevents d.c. current flowing appearing at the output terminals.

In order to stabilize the operating conditions for the stage and compensate for variations in transistor parameters, base bias current for the transistor can be derived from the voltage at the collector (see Fig. 7.31). This voltage is dependent on the collector current which, in turn, depends upon the base current. A negative feedback loop thus exists in which there is a degree of self-regulation. If the collector current increases, the collector voltage will fall and the base current will be reduced. The reduction in base current will produce a corresponding reduction in collector current to offset the original change. Conversely, if the collector current falls, the collector voltage will rise and the base current will increase. This, in turn, will produce a corresponding increase in collector current to offset the original change.

The negative feedback path in Fig. 7.31 provides feedback that involves an a.c. (signal) component as well as the d.c. bias. As a result of the a.c. feedback, there is a slight reduction in signal gain. The signal gain can be increased by removing the a.c. signal component from the feedback path so that only the d.c. bias component is present. This can be achieved with the aid of a bypass capacitor as shown in Fig. 7.32. The value of bypass capacitor, $C2$, is chosen so that the component exhibits a very low reactance at the lowest signal frequency when compared with the series base bias resistance, $R2$. The result of this potential divider arrangement is that the a.c. signal component is effectively bypassed to ground.

Figure 7.33 shows an improved form of transistor amplifier in which d.c. negative feedback is used to stabilize the stage and compensate for variations in transistor parameters, component values and temperature changes. $R1$ and $R2$ form a potential divider that determines the d.c. base potential, V_B. The base–emitter voltage (V_{BE}) is the difference between the potentials present at the base (V_B) and emitter (V_E). The potential at the emitter is governed by the emitter current (I_E). If this current increases, the emitter voltage (V_E) will increase and,

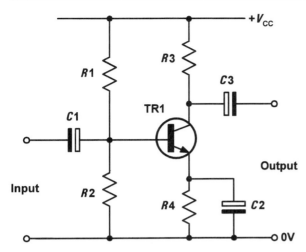

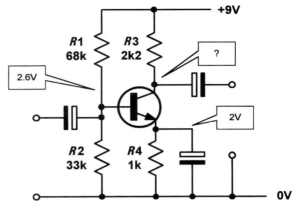

Figure 7.34

Figure 7.33 A common-emitter amplifier stage with effective bias stabilization

as a consequence V_{BE} will fall. This, in turn, produces a reduction in emitter current which largely offsets the original change. Conversely, if the emitter current decreases, the emitter voltage (V_E) will decrease and V_{BE} will increase (remember that V_B remains constant). The increase in bias results in an increase in emitter current compensating for the original change.

Example 7.8

Determine the static value of current gain and collector voltage in the circuit shown in Fig. 7.34.

Solution

Since 2 V appears across $R4$, we can determine the emitter current easily from:

$$I_E = \frac{V_E}{R4} = \frac{2\text{ V}}{1\text{ k}\Omega} = 2\text{ mA}$$

Next we should determine the base current. This is a little more difficult. The base current is derived from the potential divider formed by $R1$ and $R2$. The potential at the junction of $R1$ and $R2$ is 2.6 V hence we can determine the currents through $R1$ and $R2$, the difference between these currents will be equal to the base current.

The current in $R2$ will be given by:

$$I_{R2} = \frac{V_B}{R2} = \frac{2.6\text{ V}}{33\text{ k}\Omega} = 79\text{ }\mu\text{A}$$

The current in $R1$ will be given by:

$$I_{R1} = \frac{9\text{ V} - V_B}{R1} = \frac{6.4\text{ V}}{68\text{ k}} = 94.1\text{ }\mu\text{A}$$

Hence base current, $I_B = 94.1\text{ }\mu\text{A} - 79\text{ }\mu\text{A} = 15.1\text{ }\mu\text{A}$.

Next we can determine the collector current from:

$$I_C = I_B + I_C = 15.1\text{ }\mu\text{A} + 2\text{ mA} = 2.0151\text{ mA}$$

The current gain can then be determined from:

$$h_{FE} = \frac{I_C}{I_B} = \frac{2.0151\text{ mA}}{15.1\text{ }\mu\text{A}} = 133.45$$

Finally we can determine the collector voltage by subtracting the voltage dropped across $R3$ from the 9 V supply. The voltage dropped across $R4$ will be:

$$V_{R4} = I_C \times R4 = 2.0151\text{ mA} \times 2.2\text{ k}\Omega = 4.433\text{ V}$$

Hence collector voltage, $V_C = 9\text{ V} - 4.433\text{ V} = 4.567\text{ V}$.

Predicting amplifier performance

The a.c. performance of an amplifier stage can be predicted using a load line superimposed on the relevant set of output characteristics. For a bipolar transistor operating in common-emitter mode the required characteristics are I_C plotted against V_{CE}. One end of the load line corresponds to the supply voltage (V_{CC}) while the other end corresponds to the value of collector or drain current that would flow with the device totally saturated. In this condition:

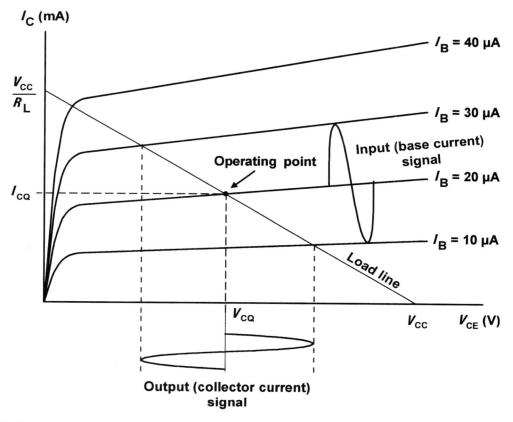

Figure 7.35   Operating point and quiescent values shown on the load line for a bipolar transistor operating in common-emitter mode.

$$I_C = \frac{V_{CC}}{R_L}$$

where R_L is the value of collector or drain load resistance.

Figure 7.35 shows a load line superimposed on a set of output characteristics for a bipolar transistor operating in common-emitter mode. The quiescent point (or operating point) is the point on the load line that corresponds to the conditions that exist when no-signal is applied to the stage. In Fig. 7.35, the base bias current is set at 20 μA so that the quiescent point effectively sits roughly halfway along the load line. This position ensures that the collector voltage can swing both positively (above) and negatively (below) its quiescent value

(V_{CQ}). The effect of superimposing an alternating base current (of 20 μA peak–peak) to the d.c. bias current (of 20 μA) can be clearly seen. The corresponding collector current signal can be determined by simply moving up and down the load line.

Example 7.9

The characteristic curves shown in Fig. 7.36 relate to a transistor operating in common-emitter mode. If the transistor is operated with $I_B = 30$ μA, a load resistor of 1.2 kΩ and an 18 V supply, determine the quiescent values of collector voltage and current (V_{CQ} and I_{CQ}). Also determine the peak–peak output voltage that would be produced by an input signal of 40 μA peak–peak.

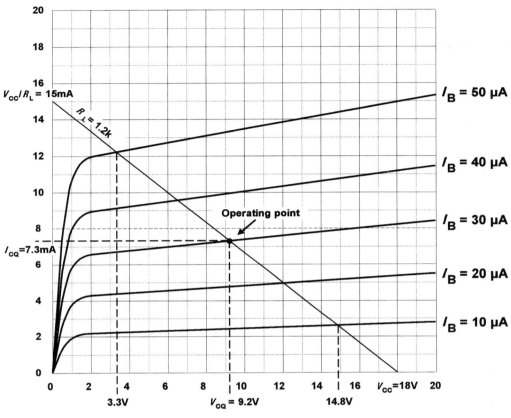

Figure 7.36

Solution

First we need to construct the load line. The two ends of the load line will correspond to V_{CC} (18 V) on the collector–emitter voltage axis and V_{CC}/R_L (18 V/1.2 kΩ or 15 mA) on the collector current axis. Next we locate the operating point (or quiescent point) from the point of intersection of the $I_B = 30$ μA characteristic and the load line.

Having located the operating point we can read off the quiescent (no-signal) values of collector–emitter voltage (V_{CQ}) and collector current (I_{CQ}). Hence:

$V_{CQ} = 9.2$ V and $I_{CQ} = 7.3$ mA

Next we can determine the maximum and minimum values of collector–emitter voltage by locat-

ing the appropriate intercept points on Fig. 7.36. Note that the maximum and minimum values of base current will be (30 μA + 20 μA) on positive peaks of the signal and (30 μA − 20 μA) on negative peaks of the signal.

The maximum and minimum values of V_{CE} are, respectively, 14.8 V and 3.3 V. Hence the output voltage swing will be (14.8 V − 3.3 V) or 11.5 V pk–pk.

Practical amplifier circuits

The simple common-emitter amplifier stage shown in Fig. 7.37 provides a modest voltage gain (80 to

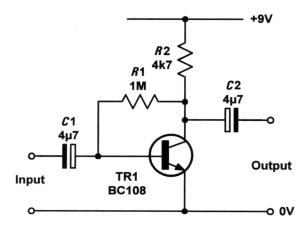

Figure 7.37 A practical common-emitter amplifier stage

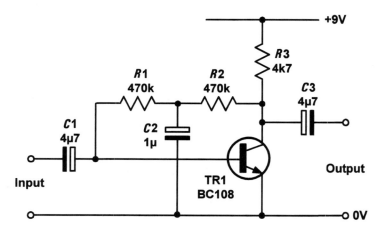

Figure 7.38 An improved common-emitter amplifier stage

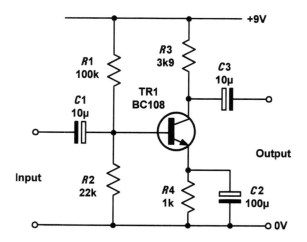

Figure 7.39 A practical common-emitter amplifier stage with bias stabilization

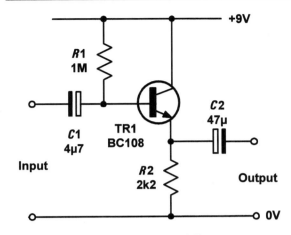

Figure 7.40   A practical emitter-follower stage

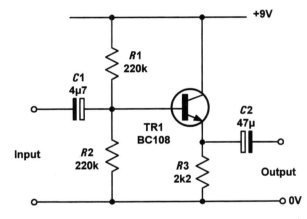

Figure 7.41   An improved emitter-follower stage

120 typical) with an input resistance of approximately 1.5 kΩ and an output resistance of around 20 kΩ. The frequency response extends from a few hertz to several hundred kilohertz.

The improved arrangement shown in Fig. 7.38 provides a voltage gain of around 150 to 200 by eliminating the signal-frequency negative feedback that occurs through $R1$ in Fig. 7.37.

Figure 7.39 shows a practical common-emitter amplifier with bias stabilization. This stage provides again of 150 to well over 200 (depending upon the current gain, h_{fe}, of the individual transistor used). The circuit will operate with supply voltages of between 6 V and 18 V.

Two practical emitter-follower circuits are shown in Figs 7.40 and 7.41. These circuits offer a voltage gain of unity (1) but are ideal for matching a high resistance source to a low resistance load. It is important to note that the input resistance varies with the load connected to the output of the circuit (it is typically in the range 50 kΩ to 150 kΩ). The input resistance can be calculated by multiplying h_{fe} by the effective resistance of $R2$ in parallel with the load connected to the output terminals.

Figure 7.41 is an improved version of Fig. 7.40 in which the base current is derived from the potential divider formed by $R1$ and $R2$. Note, however, that the input resistance is reduced since $R1$ and $R2$ effectively appear in parallel with the input. The input resistance of the stage is thus typically in the region of 40 kΩ to 70 kΩ.

Symbols introduced in this chapter

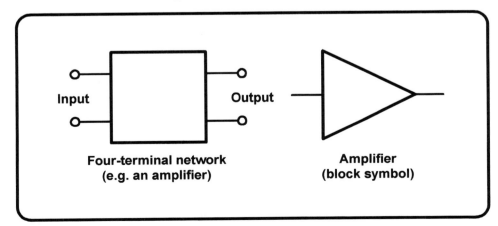

Figure 7.42

Important formulae introduced in this chapter

Voltage gain:
(page 121)

$$A_V = \frac{V_{out}}{V_{in}}$$

Current gain:
(page 121)

$$A_i = \frac{I_{out}}{I_{in}}$$

Power gain:
(page 122)

$$A_p = \frac{P_{out}}{P_{in}}$$

$$A_p = A_i \times A_v$$

Gain with negative feedback applied:
(page 128)

$$A_{VNFB} = \frac{A_V}{1 + \beta A_V}$$

Loop gain:
(page 128)

$$A_{VLOOP} = \beta A_V$$

Input resistance (hybrid equiv.):
(page 131)

$$h_i = \frac{dv_i}{di_i}$$

Reverse voltage transfer ratio (hybrid equiv.):
(page 131)

$$h_r = \frac{dv_i}{dv_o}$$

Forward current transfer ratio (hybrid equiv.):
(page 131)

$$h_f = \frac{di_o}{di_i}$$

Output conductance (hybrid equiv.):
(page 131)

$$h_o = \frac{di_o}{dv_o}$$

Input resistance (common emitter):
(page 132)

$$h_{ie} = \frac{dv_{be}}{di_b}$$

Reverse voltage transfer ratio (common emitter):
(page 132)

$$h_{re} = \frac{dv_{be}}{dv_{ce}}$$

Forward current transfer ratio (common emitter):
(page 132)

$$h_{fe} = \frac{di_c}{di_b}$$

Output conductance (common emitter):
(page 132)

$$h_{oe} = \frac{di_c}{dv_{ce}}$$

Voltage gain (common emitter, assuming h_{re} and can be neglected):
(page 135)

$$A_v = \frac{h_{fe} \times R_L}{h_{ie}}$$

Problems

7.1 The following measurements were made during a test on an amplifier:

V_{in} = 250 mV, I_{in} = 2.5 mA, V_{out} = 10 V, I_{out} = 400 mA.

Determine:

(a) the voltage gain;
(b) the current gain;
(c) the power gain;
(d) the input resistance.

7.2 An amplifier has a power gain of 25 and identical input and output resistances of 600 Ω. Determine the input voltage required to produce an output of 10 V.

7.3 Determine the mid-band voltage gain and upper and lower cut-off frequencies for the amplifier whose frequency response curve is shown in Fig. 7.43.

7.4 An amplifier with negative feedback applied

Voltage gain

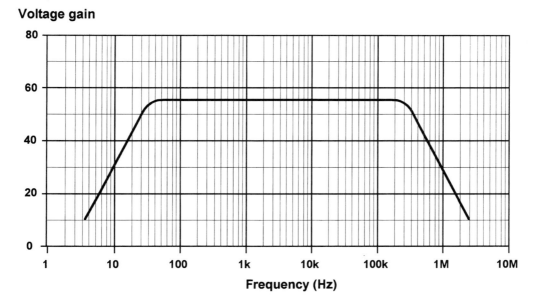

Figure 7.43

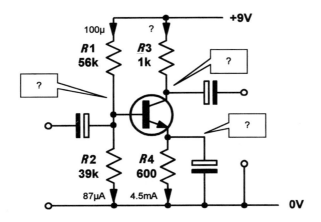

Figure 7.44

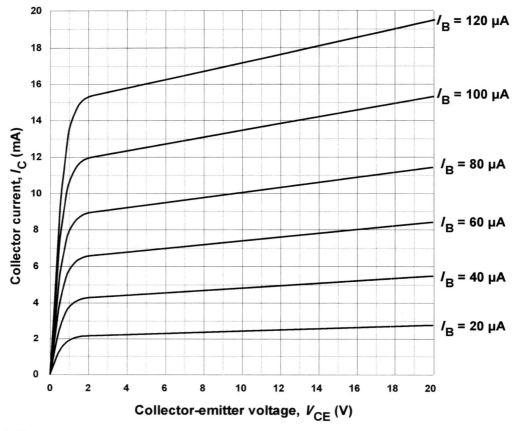

Figure 7.45

has an open-loop voltage gain of 250, and 5% of its output is fed back to the input. Determine the overall voltage gain with negative feedback applied.

7.5   An amplifier produces an open-loop gain of 180. Determine the amount of feedback required if it is to be operated with a precise voltage gain of 50.

7.6   A transistor has the following h-parameters:

$h_{ie} = 800\ \Omega$
$h_{re} = $ negligible
$h_{fe} = 120$
$h_{oe} = 50\ \mu S$.

If the transistor is to be used as the basis of a common-emitter amplifier stage with $R_L = $

12 kΩ, determine the output voltage when an input signal of 2 mV is applied.

7.7   Determine the unknown current and voltages in Fig. 7.44.

7.8   The output characteristics of a bipolar transistor are shown in Fig. 7.45. If this transistor is used in an amplifier circuit operating from a 12 V supply with a base bias current of 60 μA and a load resistor of 1 kΩ, determine the quiescent values of collector–emitter voltage and collector current. Also determine the peak–peak output voltage produced when an 80 μA peak–peak signal current is applied.

7.9   The output characteristics of a field effect transistor are shown in Fig. 7.46. If this FET is used in an amplifier circuit operating from an

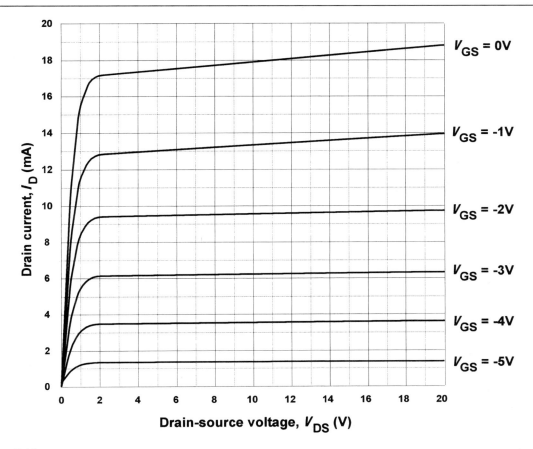

Figure 7.46

18 V supply with a gate-source bias voltage of −3 V and a load resistor of 900 Ω, determine the quiescent values of drain-source voltage and drain current. Also determine the peak–peak output voltage when an input voltage of 2 V pk–pk is applied. Also determine the voltage gain of the stage.

(Answers to these problems can be found on page 202.)

8

Operational amplifiers

This chapter introduces a highly versatile family of analogue integrated circuits. These operational amplifier 'gain blocks' offer near-ideal characteristics (i.e. virtually infinite voltage gain and input resistance coupled with low output resistance and wide bandwidth).

External components are added to operational amplifiers in order to define their function within a circuit. By adding two resistors, we can produce an amplifier having a precisely defined gain. Alternatively, with just one resistor and one capacitor we can produce an active integrating circuit. This chapter introduces the basic concepts of operational amplifiers and describes their use in a number of practical circuit applications.

Symbols and connections

The symbol for an operational amplifier is shown in Fig. 8.1. There are a few things to note about this. The operational amplifier has two input connections and one output connection. There is no direct connection to common. Furthermore, to keep circuits simple we don't always show the connections to the supply – it is often clearer to leave them out of the circuit altogether!

In Fig. 8.1, one of the inputs is marked '−' and the other is marked '+'. These polarity markings have nothing to do with the supply connections – they indicate the overall phase shift between each input and the output. The '+' sign indicates zero phase shift while the '−' sign indicates 180° phase shift. Since 180° phase shift produces an inverted (i.e. turned upside down) waveform, the '−' input

is often referred to as the **inverting input**. Similarly, the '+' input is known as the **non-inverting input**.

Most (but not all) operational amplifiers require a **symmetrical supply** (of typically ±5 V to ±15 V). This allows the output voltage to swing both positive (above 0 V) and negative (below 0 V). Figure 8.2 shows how the supply connections would appear if we decided to include them. Note that we usually have two separate supplies; a positive supply and an equal, but opposite, negative supply. The common connection to these two supplies (i.e. the 0 V rail) acts as the **common rail** in our circuit. The input and output voltages are usually measured relative to this rail. Figure 8.3 shows how the supplies are connected.

Terminology

Before we take a look at some of the characteristics of 'ideal' and 'real' operational amplifiers it is important to define some of the terms that we apply to these devices. Some of these terms (such as voltage gain, input resistance and output resistance) were introduced briefly in Chapter 7. We shall now expand on the definitions of these terms in relation to operational amplifiers.

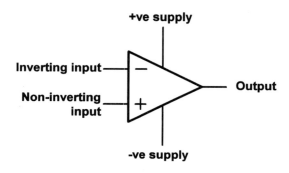

Figure 8.2 Operational amplifier with supply connections

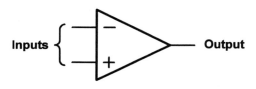

Figure 8.1 Symbol for an operational amplifier

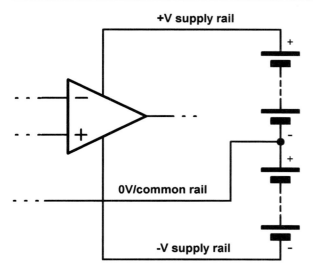

Figure 8.3 A typical operational amplifier power supply arrangement

Open-loop voltage gain

The open-loop voltage gain of an operational amplifier is defined as the ratio of output voltage to input voltage measured with no feedback applied. Open-loop voltage gain may thus be thought of as the 'internal' voltage gain of the device. In practice, this value is exceptionally high (typically greater than 100 000) but is liable to considerable variation from one device to another.

Open-loop voltage gain is the ratio of output voltage to input voltage measured without feedback applied, hence:

$$A_{VOL} = V_{out}/V_{in}$$

where A_{VOL} is the open-loop voltage gain, V_{out} and V_{in} are the output and input voltages, respectively, under open-loop conditions. In linear voltage amplifying applications, a large amount of negative feedback will normally be applied and the open-loop voltage gain can be thought of as the internal voltage gain provided by the device.

The open-loop voltage gain is often expressed in decibels (dB) rather than as a ratio (see Appendix 5). In this case:

$$A_{VOL} = 20 \log_{10}(V_{out}/V_{in})$$

Closed-loop voltage gain

The closed-loop voltage gain of an operational amplifier is defined as the ratio of output voltage to input voltage measured with a small proportion of the output fed back to the input (i.e. with feedback applied). The effect of providing negative feedback is to reduce the loop voltage gain to a value which is both predictable and manageable. Practical closed-loop voltage gains range from 1 to several thousand but note that high values of voltage gain may make unacceptable restrictions on bandwidth (see later).

Closed-loop voltage gain is the ratio of output voltage to input voltage when negative feedback is applied, hence:

$$A_{VCL} = V_{out}/V_{in}$$

where A_{VCL} is the closed-loop voltage gain, V_{out} and V_{in} are the output and input voltages, respectively, under closed-loop conditions. The closed-loop voltage gain is normally very much less than the open-loop voltage gain.

Example 8.1

An operational amplifier operating with negative feedback produces an output voltage of 2 V when supplied with an input of 400 μV. Determine the value of closed-loop voltage gain and express your answer in decibels.

Solution

Now

$$A_{VCL} = V_{out}/V_{in}$$

thus

$$A_{VCL} = 2 \text{ V}/400 \text{ μV} = 5000$$

In decibels, the value of voltage gain will be given by:

$$A_{VCL} = 20 \log_{10}(5000) = 20 \times 3.7 = 74 \text{ dB}$$

Input resistance

The input resistance of an operational amplifier is defined as the ratio of input voltage to input current expressed in ohms. It is often expedient to assume that the input of an operational amplifier is purely resistive, although this is not the case at high frequencies where shunt capacitive reactance may become significant. The input resistance of operational amplifiers is very much dependent on the semiconductor technology employed. In practice values range from about 2 MΩ for common bipolar types to over $10^{12}\,\Omega$ for FET and CMOS devices.

Input resistance is the ratio of input voltage to input current:

$$R_{in} = V_{in}/I_{in}$$

where R_{in} is the input resistance (Ω), V_{in} is the input voltage (V) and I_{in} is the input current (A). Note that we usually assume that the input of an operational amplifier is purely resistive though this may not be the case at high frequencies where shunt capacitive reactance may become significant.

Example 8.2

An operational amplifier has an input resistance of 2 MΩ. Determine the input current when an input voltage of 5 mV is present.

Solution

Since

$$R_{in} = V_{in}/I_{in}$$

$$I_{in} = V_{in}/R_{in} = 5 \text{ mV}/2 \text{ M}\Omega = 2.5 \text{ nA}$$

Output resistance

The output resistance of an operational amplifier is defined as the ratio of open-circuit output voltage to short-circuit output current expressed in ohms. Typical values of output resistance range from less than 10 kΩ to around 100 kΩ depending upon the configuration and amount of feedback employed.

Output resistance is the ratio of open-circuit output voltage to short-circuit output current, hence:

$$R_{out} = V_{out(OC)}/I_{out(SC)}$$

where R_{out} is the output resistance (Ω), $V_{out(OC)}$ is the open-circuit output voltage (V) and $I_{out(SC)}$ is the short-circuit output current (A).

Input offset voltage

An ideal operational amplifier would provide zero output voltage when 0 V is applied to its input. In practice, due to imperfect internal balance, there may be some small voltage present at the output. The voltage that must be applied differentially to the operational amplifier input in order to make the output voltage exactly zero is known as the input offset voltage.

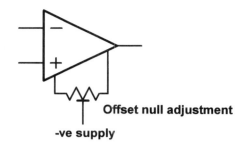

Offset null adjustment

-ve supply

Figure 8.4   Offset null facility

Offset voltage may be minimized by applying relatively large amounts of negative feedback or by using the **offset null** facility provided by a number of operational amplifier devices. Typical values of input offset voltage range from 1 mV to 15 mV. Where a.c., rather than d.c., coupling is employed, offset voltage is not normally a problem and can be happily ignored.

The input offset voltage is the voltage which, when applied at the input, provides an output voltage of exactly zero. Similarly, the input offset current is the current which, when applied at the input, provides an output voltage of exactly zero. (Note that, due to imperfect balance and very high internal gain a small output voltage may appear with no input present.) Offset may be minimized by applying large amounts of negative feedback or by using the offset null facility provided by a number of operational amplifiers (see Fig. 8.4).

Full-power bandwidth

The full-power bandwidth is equivalent to the frequency at which the maximum undistorted peak output voltage swing falls to 0.707 of its low frequency (d.c.) value (the sinusoidal input voltage remaining constant). Typical full-power bandwidths range from 10 kHz to over 1 MHz for some high-speed devices.

Slew rate

Slew rate is the rate of change of output voltage with time, when a rectangular step input voltage is applied. Slew rate is measured in V/s (or V/µs) and typical values range from 0.2 V/µs to over 20 V/µs. Slew rate imposes a limitation on circuits in which large amplitude pulses rather than small amplitude sinusoidal signals are likely to be encountered.

The slew-rate of an operational amplifier is the rate of change of output voltage with time in response to a perfect step-function input. Hence:

slew-rate = dV_{out}/dt

where dV_{out} is the change in output voltage (V) and dt is the corresponding interval of time (s).

Common-mode rejection ratio

Common-mode rejection ratio is a measure of an operational amplifier's ability to ignore signals simultaneously present on both inputs (i.e. 'common-mode' signals) in preference to signals applied differentially. Common-mode rejection ratio is defined as the ratio of differential voltage gain to common-mode voltage gain.

Common-mode rejection ratio is usually specified in decibels (see Appendix 5) and typical values range from 80 dB to 110 dB. Common-mode rejection ratio (CMRR) is the ratio of differential voltage gain to common-mode voltage gain (usually expressed in dB). Hence:

$$CMRR = 20 \log_{10} \frac{(A_{VOL(DM)})}{(A_{VOL(CM)})}$$

where $A_{VOL(DM)}$ is the open-loop voltage gain in differential mode (equal to A_{VOL}) and $A_{VOL(CM)}$ is the open-loop voltage gain in common mode (i.e. signal applied with both inputs connected together). CMRR is thus a measure of an operational amplifier's ability to reject signals (e.g. noise) which are simultaneously present on both inputs.

Example 8.3

With no feedback applied, an operational amplifier produces an output of 10 V from a differential input of 50 μV. With the inputs shorted together, the same operational amplifier produces an output of 500 mV when an input of 2 V is present. Determine the value of common-mode rejection ratio.

Solution

First we need to find the values of differential and common-mode voltage gain ($A_{VOL(DM)}$ and $A_{VOL(CM)}$):

$A_{VOL(DM)} = V_{out(DM)}/V_{in(DM)} = 10$ V/50 μV
$= 200\ 000$

and

$A_{VOL(CM)} = V_{out(CM)}/V_{in(CM)} = 500$ mV/2 V $= 0.25$

Now

$$CMRR = 20 \log_{10} \frac{(A_{VOL(DM)})}{(A_{VOL(CM)})}$$
$$= 20 \log_{10} (200\ 000/0.25)$$

thus

$CMRR = 20 \times \log_{10} (800\ 000) = 118$ dB

Maximum output voltage swing

The maximum output voltage swing produced by an operational amplifier is the maximum range of output voltages that the device can produce without distortion. Normally these will be symmetrical about 0 V and within a volt or so of the supply voltage rails (both positive and negative).

Operational amplifier characteristics

Having defined the terminology applied to operational amplifiers we shall now consider the characteristics of an 'ideal' operational amplifier. The desirable characteristics for an operational amplifier are summarized below:

(a) The open-loop voltage gain should be very high (ideally infinite).
(b) The input resistance should be very high (ideally infinite).
(c) The output resistance should be very low (ideally zero).
(d) Full-power bandwidth should be as wide as possible.
(e) Slew-rate should be as large as possible.
(f) Input offset should be as small as possible.
(g) Common-mode rejection ratio should be as large as possible.

The characteristics of modern IC operational amplifiers come very close to those of an 'ideal' operational amplifier (see Table 8.1).

Gain and bandwidth

It is important to note that, since the product of gain and bandwidth is a constant for any particular operational amplifier, an increase in gain can only

Table 8.1   Characteristics of ideal and real operational amplifiers

Parameter	Ideal	Real
Voltage gain	Infinite	100 000
Input resistance	Infinite	100 MΩ
Output resistance	Zero	20 Ω
Bandwidth	Infinite	2 MHz

Table 8.2   Table showing the relationship between voltage gain and bandwidth for Fig. 8.5

Voltage gain (A_V)	Bandwidth
1	d.c. to 1 MHz
10	d.c. to 100 kHz
100	d.c. to 10 kHz
1 000	d.c. to 1 kHz
10 000	d.c. to 100 Hz
100 000	d.c. to 10 Hz

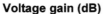

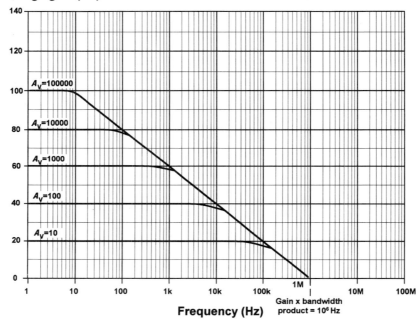

Figure 8.5   Gain plotted against bandwidth for a typical operational amplifier

be achieved at the expense of bandwidth, and vice versa.

Figure 8.5 shows the relationship between voltage gain and bandwidth for a typical operational amplifier (note that axes use logarithmic rather than linear scales). The open-loop voltage gain (i.e. that obtained with no feedback applied) is 100 000 (or 100 dB) and the bandwidth obtained in this condition is a mere 10 Hz. The effect of applying increasing amounts of negative feedback (and consequently reducing the gain to a more manageable amount) is that the bandwidth increases in direct proportion.

Frequency response curves have been added to Fig. 8.5 to show the effect on the bandwidth of making the closed-loop gains equal to 10 000, 1000, 100, and 10. Table 8.2 summarizes these results. You should also note that the (gain × bandwidth) product for this amplifier is 1×10^6 Hz (i.e. 1 MHz).

Example 8.4

The open-loop frequency response of an operational amplifier is shown in Fig. 8.6. Determine the bandwidth of the amplifier if the closed-loop voltage gain is set at 46 dB.

Voltage gain (dB)

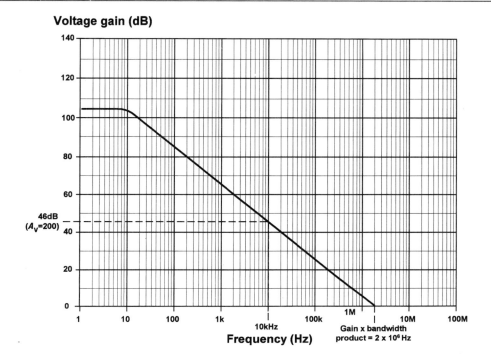

Figure 8.6

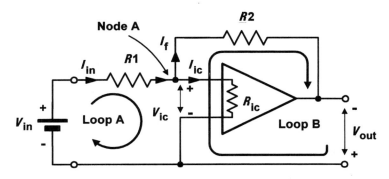

Figure 8.7   Inverting amplifier showing currents and voltages

Solution

We can determine the bandwidth of the amplifier when the closed-loop voltage gain is set to 46 dB by constructing a line and noting the intercept point on the response curve (as shown in Fig. 8.6). This shows that the bandwidth will be 10 kHz (note that, for this operational amplifier, the (gain × bandwidth) product is 2×10^6 Hz (or 2 MHz).

Inverting operational amplifier stage

Figure 8.7 shows the circuit of an operational amplifier with negative feedback applied. For the sake of our explanation we will assume that the operational amplifier is 'ideal'. Now consider what happens when a small positive input voltage is applied. This voltage (V_{in}) produces a current (I_{in})

flowing in the input resistor in the direction shown in Fig. 8.7.

Since the operational amplifier is 'ideal' we will assume that:

(a) the input resistance (i.e. the resistance that appears between the inverting and non-inverting input terminals, R_{ic}) is infinite;
(b) the open-loop voltage gain (i.e. the ratio of V_{out} to V_{in} with no feedback applied) is infinite.

As a consequence of (a) and (b):

(i) the voltage appearing between the inverting and non-inverting inputs (V_{ic}) will be zero; and
(ii) the current flowing into the chip (I_{ic}) will be zero (recall that $I_{ic} = V_{ic}/R_{ic}$ and R_{ic} is infinite).

Applying Kirchhoff's current law at node A gives:

$$I_{in} = I_{ic} + I_f \text{ but } I_{ic} = 0 \text{ thus } I_{in} = I_f \qquad (1)$$

(this shows that the current in the feedback resistor, $R2$, is the same as the input current, I_{in}).

Applying Kirchhoff's voltage law to loop A gives:

$$V_{in} = (I_{in} \times R1) + V_{ic} \text{ but } V_{ic} = 0$$
$$\text{thus } V_{in} = I_{in} \times R1 \qquad (2)$$

Applying Kirchhoff's voltage law to loop B gives:

$$V_{out} = -V_{ic} + (I_f \times R2) \text{ but } V_{ic} = 0$$
$$\text{thus } V_{out} = I_f \times R2 \qquad (3)$$

Combining (1) and (3) gives:

$$V_{out} = I_{in} \times R2 \qquad (4)$$

The voltage gain of the stage is given by:

$$A_V = \frac{V_{out}}{V_{in}} \qquad (5)$$

Combining (4) and (2) with (5) gives:

$$A_V = \frac{I_{in} \times R2}{I_{in} \times R1} = \frac{R2}{R1}$$

Improving symmetry

To preserve symmetry and minimize offset voltage, a third resistor is often included in series with the non-inverting input (see Fig. 8.8). The value of this resistor should be equivalent to the parallel combination of $R1$ and $R2$. Hence:

$$R3 = \frac{R1 \times R2}{R1 + R2}$$

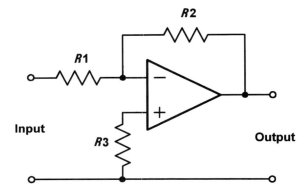

Figure 8.8   Improving the symmetry of an inverting amplifier

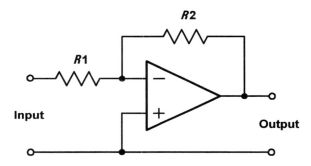

Figure 8.9   Basic inverting amplifier circuit

Operational amplifier circuit configurations

The three basic configurations for operational voltage amplifiers are shown in Figs 8.9, 8.10 and 8.11. Supply rails have been omitted from these diagrams for clarity but are assumed to be symmetrical about 0 V, as in Fig. 8.3. All three of these basic arrangements are d.c. coupled and their characteristics are summarized in Table 8.3.

Tailoring the frequency response

All of the amplifier circuits described previously have used direct coupling and thus have frequency response characteristics which extend to d.c. This, of course, is undesirable for many applications, particularly where a wanted a.c. signal may be superimposed on an unwanted d.c. voltage level. In such cases a capacitor of appropriate value may be inserted in series with the input, as shown in Fig. 8.12. The value of this capacitor should be chosen

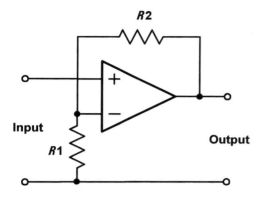

Figure 8.10 Basic non-inverting amplifier circuit

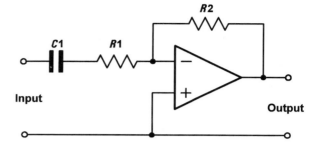

Figure 8.12 Inverting amplifier with a.c. input coupling

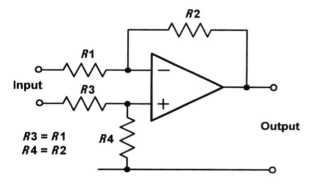

Figure 8.11 Basic differential amplifier circuit

so that its reactance is very much smaller than the input resistance at the lower applied input frequency. The effect of the capacitor on an amplifier's frequency response is shown in Fig. 8.13.

We can also use a capacitor to restrict the upper frequency response of an amplifier. This time, the capacitor is connected as part of the feedback path. Indeed, by selecting appropriate values of capacitor, the frequency response of an inverting operational voltage amplifier may be very easily tailored

to suit individual requirements (see Figs 8.14 and 8.15). The lower cut-off frequency is determined by the value of the input capacitance, $C1$, and input resistance, $R1$. The lower cut-off frequency is given by:

$$f2 = \frac{1}{2\pi C1R1} = \frac{0.159}{C1R1} \quad \text{Hz}$$

where $C1$ is in farads and $R1$ is in ohms.

Provided the upper frequency response it not limited by the gain × bandwidth product, the upper cut-off frequency will be determined by the feedback capacitance, $C2$, and feedback resistance, $R2$, such that:

$$f2 = \frac{2}{2\pi C2R2} = \frac{0.159}{C2R2} \quad \text{Hz}$$

where $C2$ is in farads and $R2$ is in ohms.

Example 8.5

An audio voltage amplifier is to operate according to the following specification:

Voltage gain	100
Input resistance (at mid-band)	10 kΩ
Phase shift (at mid-band)	180°
Lower cut-off frequency	250 Hz
Upper cut-off frequency	15 kHz

Table 8.3 Characteristics of the operational amplifier circuits shown in Figs 8.9, 8.10 and 8.11

Amplifier type	Input resistance	Voltage gain	Phase shift
Inverting amplifier (Fig. 8.9)	$R1$	$R2/R1$	180°
Non-inverting amplifier (Fig. 8.10)	$R_{in} \times \dfrac{A_{OL}^*}{1 + (R2/R1)}$	$1 + (R2/R1)$	0°
Differential amplifier (Fig. 8.11)	$2R1$	$R2/R1$	180°

* where R_{in} is the input resistance of the operational amplifier, and A_{OL} is the open-loop voltage gain of the operational amplifier.

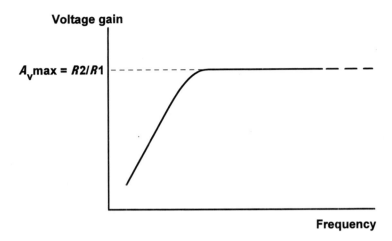

Figure 8.13   Effect of $C1$ on the frequency response of the circuit shown in Fig. 8.12

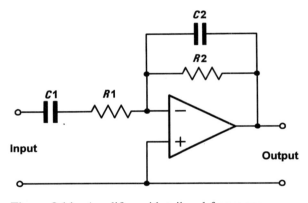

Figure 8.14   Amplifier with tailored frequency response

Devise a circuit to satisfy the above specification using an operational amplifier.

Solution

To make things a little easier, we can break the problem down into manageable parts.

Since 180° phase shift is required, we shall base our circuit on a single operational amplifier with capacitors to define the upper and lower cut-off frequencies, as shown in Fig. 8.15.

The nominal input resistance is the same as the value for $R1$. Thus:

$R1 = 10 \text{ k}\Omega$

To determine the value of $R2$ we can make use of the formula for mid-band voltage gain:

$A_V = R2/R1$

thus

$R2 = A_V \times R1 = 100 \times 10 \text{ k}\Omega = 100 \text{ k}\Omega$

To determine the value of $C1$ we will use the formula for the low frequency cut-off:

$$f2 = \frac{1}{2\pi C1R1} = \frac{0.159}{C1R1} \quad \text{Hz}$$

thus

$$C1 = \frac{0.159}{f1R1} = \frac{0.159}{250 \times 10\ 000}$$

$$= 63.6 \times 10^{-9} \text{ F} = 63.6 \text{ nF}$$

Finally, to determine the value of $C2$ we will use the formula for high frequency cut-off:

$$f2 = \frac{1}{2\pi C2R2} = \frac{0.159}{C2R2} \quad \text{Hz}$$

thus

$$C2 = \frac{0.159}{f2R2} = \frac{0.159}{15\ 000 \times 100\ 000}$$

$$= 0.106 \times 10^{-9} \text{ F} = 106 \text{ pF}$$

Operational amplifier types and packages

Operational amplifiers are available packaged singly, in pairs (dual types), or in fours (quad types). The majority of operational amplifiers are supplied in dual-in-line (DIL) packages but TO5 packages

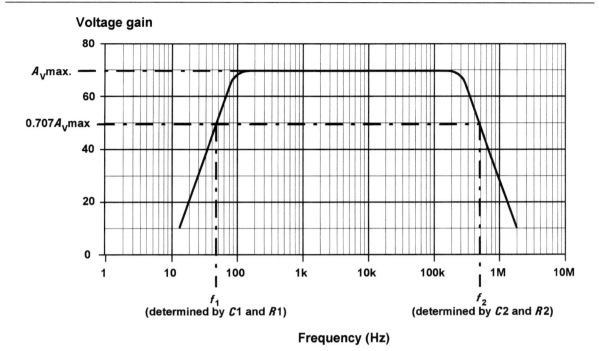

Voltage gain

Figure 8.15 Frequency response of the circuit shown in Fig. 8.14

are also used. The 081, for example, is a single general-purpose BIFET operational amplifier housed in an 8-pin DIL package. This device is also available in dual (082) and quad (084) forms (see Fig. 8.16). Table 8.4 summarizes the data for some common operational amplifier types.

Important formulae introduced in this chapter

Open-loop voltage gain:
(page 147)

$A_{VOL} = V_{out}/V_{in}$

$A_{VOL} = 20 \log_{10}(V_{out}/V_{in})$

Input resistance:
(page 147)

$R_{in} = V_{in}/I_{in}$

Output resistance:
(page 148)

$R_{out} = V_{out(OC)}/I_{out(SC)}$

Slew-rate:
(page 148)

$S = dV_{out}/dt$

Common-mode rejection ratio:
(page 149)

$$CMRR = 20 \log_{10} \frac{(A_{VOL\,(DM)})}{(A_{VOL\,(CM)})}$$

Voltage gain of an inverting amplifier:
(page 152)

$$A_V = \frac{R_2}{R_1}$$

Voltage gain of a non-inverting amplifier:
(page 152)

$$A_V = 1 + \frac{R_2}{R_1}$$

Upper cut-off frequency:
(page 153)

$$f2 = \frac{1}{2\pi C_2 R_2}$$

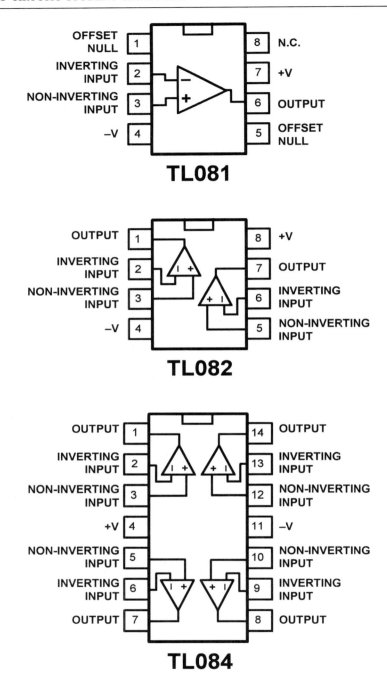

Figure 8.16 Single, dual and quad integrated circuit DIL packages

Table 8.4   Characteristics of some common operational amplifiers

Device	Type	Supply voltage range (V)	Open-loop voltage gain (dB)	Input bias current	Slew rate (V/μs)
AD548	Bipolar	4.5 to 18	100 min.	0.01 nA	1.8
AD711	FET	4.5 to 18	100	25 pA	20
CA3140	CMOS	4 to 36	100	5 pA	9
		(or ±2 to ±18)			
LF347	FET	5 to 18	110	50 pA	13
LM301	Bipolar	5 to 18	88	70 nA	0.4
LM348	Bipolar	10 to 18	96	30 nA	0.6
TL071	FET	3 to 18	106	30 pA	13
741	Bipolar	5 to 18	106	80 nA	0.5

Symbols introduced in this chapter

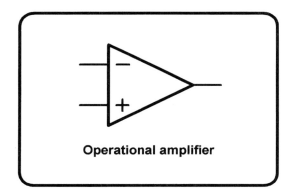

Operational amplifier

Figure 8.17

Lower cut-off frequency:
(page 153)

$$f1 = \frac{1}{2\pi C_{IN} R_{IN}}$$

Problems

8.1   An operational amplifier has an open-loop voltage gain of 100 dB. If the inverting input is held at 0 V and the non-inverting input is connected to a voltage source of 0.1 mV, determine the output voltage.

8.2   The open-loop frequency response of an operational amplifier is shown in Fig. 8.18. Determine:

(a) the voltage gain for a bandwidth of 60 kHz;
(b) the bandwidth for a voltage gain of 20 dB.

8.3   An inverting operational amplifier is required to have a voltage gain of 40 and an input resistance of 5 kΩ. Determine the value of feedback resistance required.

8.4   An operational amplifier is connected in differential mode in the circuit shown in Fig. 8.19. If $R2 = R4 = 20$ kΩ and $R1 = R3 = 1$ kΩ, determine the output voltage when the following values of input voltages (measured with respect to common) are applied:

	V1	V2
(a)	0 V	+50 mV
(b)	+50 mV	0 V
(c)	+50 mV	+50 mV
(d)	+50 mV	−50 mV
(e)	−50 mV	+50 mV
(f)	−50 mV	−50 mV

8.5   An operational amplifier has a gain × bandwidth product of 2×10^5. Estimate the bandwidth when the device is configured for closed-loop voltage gains of:

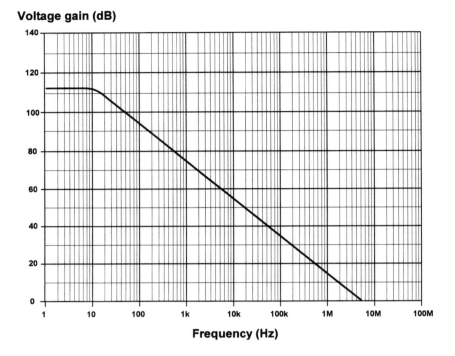

Figure 8.18

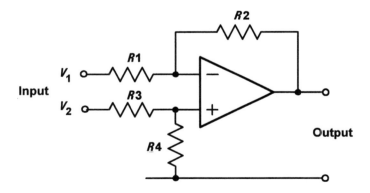

Figure 8.19

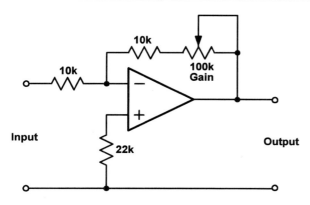

Figure 8.20

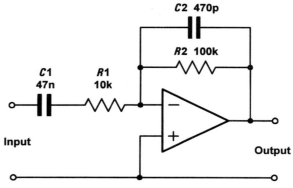

Figure 8.21

    (a) 50; and
    (b) 2000.

8.6   Determine the maximum and minimum voltage gain produced by the circuit shown in Fig. 8.20.

8.7   Determine mid-band voltage gain and upper and lower cut-off frequencies for the amplifier stage shown in Fig. 8.21.

8.8   An audio voltage amplifier is to operate according to the following specification:

Voltage gain	50
Input resistance (at mid-band)	5 kΩ
Mid-band phase shift:	180°
Lower cut-off frequency	10 Hz
Upper cut-off frequency	20 kHz

Devise a circuit to satisfy the above specification using an operational amplifier.

(Answers to these problems will be found on page 202.)

9

Oscillators

This chapter describes circuits that generate sine wave, square wave, and triangular waveforms. These oscillator circuits form the basis of clocks and timing arrangements as well as signal and function generators.

In Chapter 7, we showed how negative feedback can be applied to an amplifier to form the basis of a stage which has a precisely controlled gain. An alternative form of feedback, where the output is fed back in such a way as to reinforce the input (rather than to subtract from it), is known as **positive feedback**.

Figure 9.1 shows the block diagram of an amplifier stage with positive feedback applied. Note that the amplifier provides a phase shift of 180° and the feedback network provides a further 180°. Thus the overall phase shift is 0°. The overall voltage gain is given by:

$$\text{overall gain} = \frac{V_{\text{out}}}{V_{\text{in}}}$$

Now

$$V'_{\text{in}} = V_{\text{in}} + \beta V_{\text{out}} \quad \text{(by applying Kirchhoff's voltage law)}$$

thus

$$V_{\text{in}} = V'_{\text{in}} - \beta V_{\text{out}}$$

and

$$V_{\text{out}} = A_{\text{V}} \times V'_{\text{in}} \quad (A_{\text{V}} \text{ is the } \textbf{internal} \text{ gain of the amplifier})$$

Hence,

$$\text{overall gain} = \frac{A_{\text{V}} \times V'_{\text{in}}}{V'_{\text{in}} - \beta V_{\text{out}}} = \frac{A_{\text{V}} \times V'_{\text{in}}}{V'_{\text{in}} - \beta(A_{\text{V}} \times V'_{\text{in}})}$$

Thus

$$\text{overall gain} = \frac{A_{\text{V}}}{1 - \beta A_{\text{V}}}$$

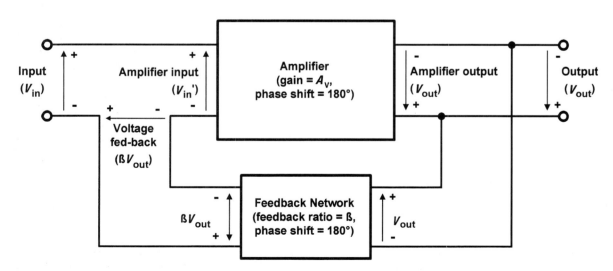

Figure 9.1 Amplifier with positive feedback applied

Now consider what will happen when the **loop gain** (βA_V) approaches unity. The denominator $(1 - \beta A_V)$ will become close to zero. This will have the effect of increasing the overall gain, i.e. the overall gain with positive feedback applied will be greater than the gain without feedback.

It is worth illustrating this difficult concept using some practical figures. Assume that you have an amplifier with a gain of 9 and one-tenth of the output is fed back to the input (i.e. $\beta = 0.1$). In this case the loop gain ($\beta \times A_V$) is 0.9.

With negative feedback applied (see Chapter 7) the overall voltage gain will fall to:

$$\frac{A_V}{1 + \beta A_V} = \frac{9}{1 + 0.1 \times 9} = \frac{9}{1 + 0.9} = \frac{9}{1.9} = 4.7$$

With positive feedback applied the overall voltage gain will be:

$$\frac{A_V}{1 - \beta A_V} = \frac{9}{1 - 0.1 \times 9} = \frac{9}{1 - 0.9} = \frac{9}{0.1} = 90$$

Now assume that you have an amplifier with a gain of 10 and, once again, one-tenth of the output is fed back to the input (i.e. $\beta = 0.1$). In this example the loop gain ($\beta \times A_V$) is exactly 1.

With negative feedback applied (see Chapter 7) the overall voltage gain will fall to:

$$\frac{A_V}{1 + \beta A_V} = \frac{10}{1 + 0.1 \times 10} = \frac{10}{1 + 1} = \frac{10}{2} = 5$$

With positive feedback applied the overall voltage gain will be:

$$\frac{A_V}{1 - \beta A_V} = \frac{10}{1 - 0.1 \times 10} = \frac{10}{1 - 1} = \frac{10}{0} = \text{infinity}$$

This simple example shows that a loop gain of unity (or larger) will result in infinite gain and an amplifier which is unstable. In fact, the amplifier will **oscillate** since any disturbance will be amplified and result in an output. Clearly, as far as an amplifier is concerned, positive feedback may have an undesirable effect – instead of reducing the overall gain the effect is that of reinforcing any signal present and the output can build up into continuous oscillation if the loop gain is 1 or greater. To put this another way, oscillator circuits can simply be thought of as amplifiers that generate an output signal without the need for an input!

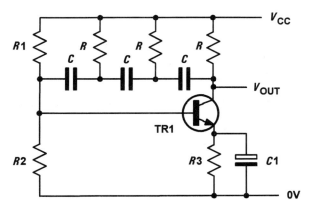

Figure 9.2 Sine wave oscillator based on a three-stage C–R ladder network

Conditions for oscillation

From the foregoing we can deduce that the conditions for oscillation are:

(a) the feedback must be positive (i.e. the signal fed back must arrive back in-phase with the signal at the input);
(b) the overall loop voltage gain must be greater than 1 (i.e. the amplifier's gain must be sufficient to overcome the losses associated with any frequency selective feedback network).

Hence, to create an oscillator we simply need an amplifier with sufficient gain to overcome the losses of the network that provide positive feedback. Assuming that the amplifier provides 180° phase shift, the frequency of oscillation will be that at which there is 180° phase shift in the feedback network. A number of circuits can be used to provide 180° phase shift, one of the simplest being a three-stage C–R ladder network. Alternatively, if the amplifier produces 0° phase shift, the circuit will oscillate at the frequency at which the feedback network produces 0° phase shift. In both cases, the essential point is that the feedback should be positive so that the output signal arrives back at the input in such a sense as to reinforce the original signal.

Ladder network oscillator

A simple phase-shift oscillator based on a three-stage C–R ladder network is shown in Fig. 9.2. TR1 operates as a conventional common-emitter amplifier stage with $R1$ and $R2$ providing base bias potential and $R3$ and $C1$ providing emitter stabilization. The total phase shift provided by the $C–R$ ladder network (connected between collector and base) is 180° at the frequency of oscillation. The transistor provides the other 180° phase shift in order to realize an overall phase shift of 360° or 0°.

The frequency of oscillation of the circuit shown in Fig. 9.2 is given by:

$$f = \frac{1}{2\pi\sqrt{6}CR}$$

The loss associated with the ladder network is 29, thus the amplifier must provide a gain of at least 29 in order for the circuit to oscillate.

Example 9.1

Determine the frequency of oscillation of a ladder network oscillator where $C = 10$ nF and $R = 10$ kΩ.

Solution

Using $f = 1/2\pi\sqrt{6}CR$ gives

$$f = \frac{1}{6.28 \times 2.45 \times 10 \times 10^{-9} \times 10 \times 10^3} \text{ Hz}$$

thus

$$f = \frac{1}{6.28 \times 2.45 \times 10^{-4}} = \frac{10^4}{15.386} = 647 \text{ Hz}$$

Wien bridge oscillator

An alternative approach to providing the phase shift required is the use of a Wien bridge network (Fig. 9.3). Like the $C–R$ ladder, this network provides a phase shift which varies with frequency. The input signal is applied to A and B while the output is taken from C and D. At one particular frequency, the phase shift produced by the network will be exactly zero (i.e. the input and output signals will be in-phase). If we connect the network to an amplifier producing 0° phase shift which has sufficient gain to overcome the losses of the Wien bridge, oscillation will result.

The minimum amplifier gain required to sustain oscillation is given by:

$$A_V = 1 + C1/C2 + R2/R1$$

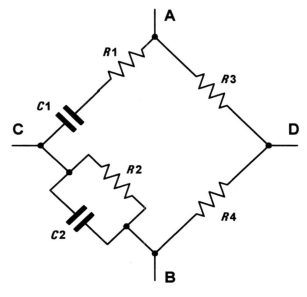

Figure 9.3   A Wien bridge network

and the frequency at which the phase shift will be zero is given by:

$$f = \frac{1}{2\pi\sqrt{(C1C2R1C2)}}$$

In practice, we normally make $R1 = R2$ and $C1 = C2$ hence:

$$A_V = 1 + C/C + R/R = 1 + 1 + 1 = 3$$

and the frequency at which the phase shift will be zero is given by:

$$f = \frac{1}{2\pi\sqrt{(RRCC)}} = \frac{1}{2\pi CR}$$

where $R = R1 = R2$ and $C = C1 = C2$.

Example 9.2

Figure 9.4 shows the circuit of a Wien bridge oscillator based on an operational amplifier. If $C1 = C2 = 100$ nF, determine the output frequencies produced by this arrangement (a) when $R1 = R2 = 1$ kΩ and (b) when $R1 = R2 = 6$ kΩ.

Solution

(a) When $R1 = R2 = 1$ kΩ

$$f = \frac{1}{2\pi CR}$$

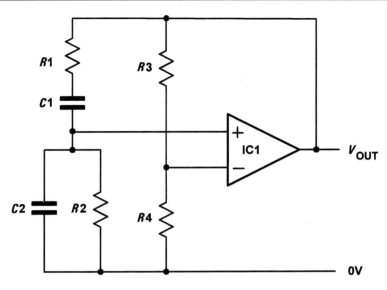

Figure 9.4 Sine wave oscillator based on a Wien bridge network (see Example 3.2)

where $R = R1 = R2$ and $C = C1 = C2$. Thus

$$f = \frac{1}{6.28 \times 100 \times 10^{-9} \times 1 \times 10^3}$$

$$= \frac{1}{6.28 \times 10^{-4}}$$

or

$$f = \frac{10\,000}{6.28} = 1592 \text{ Hz} = 1.592 \text{ kHz}$$

(b) When $R1 = R2 = 6 \text{ k}\Omega$

$$f = \frac{1}{2\pi CR} = \frac{1}{6.28 \times 100 \times 10^{-9} \times 6 \times 10^3}$$

$$= \frac{1}{37.68 \times 10^{-4}}$$

or

$$f = \frac{10\,000}{37.68} = 265.4 \text{ Hz}$$

Multivibrators

There are many occasions when we require a square wave output from an oscillator rather than a sine wave output. Multivibrators are a family of oscillator circuits that produce output waveforms consisting of one or more rectangular pulses. The term 'multivibrator' simply originates from the fact that this type of waveform is rich in harmonics (i.e. 'multiple vibrations').

Multivibrators use regenerative (positive) feedback; the active devices present within the oscillator circuit being operated as switches, being alternately cut off and driven into saturation.

The principal types of multivibrator are:

(a) **astable multivibrators** that provide a continuous train of pulses (these are sometimes also referred to as **free-running multivibrators**);

(b) **monostable multivibrators** that produce a single output pulse (they have one stable state and are thus sometimes also referred to as **one-shot** circuits);

(c) **bistable multivibrators** that have two stable states and require a trigger pulse or control signal to change from one state to another.

The astable multivibrator

Figure 9.5 shows a classic form of astable multivibrator based on two transistors. Figure 9.6 shows how this circuit can be redrawn in an arrangement that more closely resembles a two-stage common-emitter amplifier with its output connected back to its input.

In Fig. 9.5, the values of the base resistors, $R3$ and $R4$, are such that the sufficient base current will

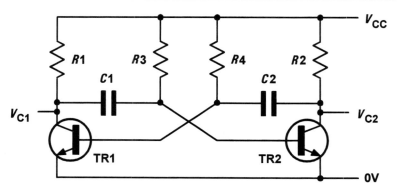

Figure 9.5   Astable multivibrator using bipolar transistors

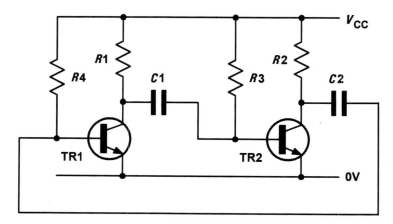

Figure 9.6   Circuit of Fig. 9.5 redrawn to show the two common-emitter stages

be available to completely saturate the respective transistor. The values of the collector load resistors, $R1$ and $R2$, are very much smaller than $R3$ and $R4$. When power is first applied to the circuit, assume that TR2 saturates before TR1 when the power is first applied (in practice one transistor would always saturate before the other due to variations in component tolerances and transistor parameters).

As TR2 saturates, its collector voltage will fall rapidly from $+V_{CC}$ to 0 V. This drop in voltage will be transferred to the base of TR1 via $C2$. This negative going voltage will ensure that TR1 is initially placed in the non-conducting state. As long as TR1 remains cut off, TR2 will continue to be saturated. During this time, $C2$ will charge via $R4$ and TR1's base voltage will rise exponentially from $-V_{CC}$ towards $+V_{CC}$. However, TR1's base voltage will not rise much above 0 V because, as soon as it reaches $+0.7$ V (sufficient to cause base current

to flow) TR1 will begin to conduct. As TR1 begins to turn on, its collector voltage will rapidly fall from $+V_{CC}$ to 0 V. This fall in voltage is transferred to the base of TR2 via $C1$ and, as a consequence, TR2 will turn off. $C1$ will then charge via $R3$ and TR2's base voltage will rise exponentially from $-V_{CC}$ towards $+V_{CC}$. As before, TR2's base voltage will not rise much above 0 V because, as soon as it reaches $+0.7$ V (sufficient to cause base current to flow), TR2 will start to conduct. The cycle is then repeated indefinitely.

The time for which the collector voltage of TR2 is low and TR1 is high ($T1$) will be determined by the time constant, $R4 \times C2$. Similarly, the time for which the collector voltage of TR1 is low and TR2 is high ($T2$) will be determined by the time constant, $R3 \times C1$.

The following approximate relationships apply:

$$T1 = 0.7C2R4 \qquad \text{and} \qquad T2 = 0.7C1R3$$

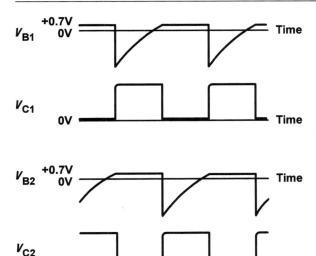

Figure 9.7 Waveforms for the transistor astable multivibrator

Since one complete cycle of the output occurs in a time, $T = T1 + T2$, the periodic time of the output is given by:

$$T = 0.7 \, (C2R4 + C1R3)$$

Finally, we often require a symmetrical 'square wave' output where $T1 = T2$. To obtain such an output, we should make $R3 = R4$ and $C1 = C2$, in which case the periodic time of the output will be given by:

$$T = 1.4CR$$

where $C = C1 = C2$ and $R = R3 = R4$.

Waveforms for the astable oscillator are shown in Fig. 9.7.

Example 9.3

The astable multivibrator in Fig. 9.4 is required to produce a square wave output at 1 kHz. Determine suitable values for $R3$ and $R4$ if $C1$ and $C2$ are both 10 nF.

Solution

Since a square wave is required and $C1$ and $C2$ have identical values, $R3$ must be made equal to $R4$. Now:

$$T = \frac{1}{f} = \frac{1}{1 \times 10^3} = 1 \times 10^{-3} \text{ s}$$

Re-arranging $T = 1.4CR$ to make R the subject gives:

$$R = \frac{T}{1.4C} = \frac{1 \times 10^{-3}}{1.4 \times 10 \times 10^{-9}} = \frac{1 \times 10^6}{14}$$
$$= 0.071 \times 10^6 \text{ } \Omega \text{ or } 71.4 \text{ k}\Omega$$

Other forms of astable oscillator

Figure 9.8 shows the circuit diagram of an alternative form of astable oscillator which produces a triangular output waveform. Operational amplifier IC1 forms an integrating stage while IC2 is

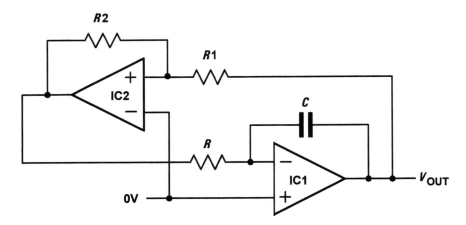

Figure 9.8 Astable oscillator using operational amplifiers

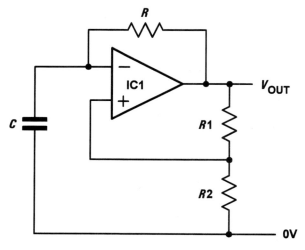

Figure 9.9   Single-stage astable oscillator using an operational amplifier

connected with positive feedback to ensure that oscillation takes place.

Assume that the output from IC2 is initially at, or near, $+V_{CC}$ and capacitor, C, is uncharged. The voltage at the output of IC2 will be passed, via R, to IC1. Capacitor, C, will start to charge and the output voltage of IC1 will begin to fall. Eventually, the output voltage will have fallen to a value that causes the polarity of the voltage at the non-inverting input of IC2 to change from positive to negative. At this point, the output of IC2 will rapidly fall to $-V_{CC}$. Again, this voltage will be passed, via R, to IC1. Capacitor, C, will then start to charge in the other direction and the output voltage of IC1 will begin to rise. Eventually, the output voltage will have risen to a value that causes the polarity of the non-inverting input of IC2 to revert to its original (positive) state and the cycle will continue indefinitely.

The upper threshold voltage (i.e. the maximum positive value for V_{out}) will be given by:

$$V_{UT} = V_{CC} \times (R1/R2)$$

The lower threshold voltage (i.e. the maximum negative value for V_{out}) will be given by:

$$V_{LT} = -V_{CC} \times (R1/R2)$$

Single-stage astable oscillator

A simple form of astable oscillator producing a square wave output can be built using just one operational amplifier, as shown in Fig. 9.9. This circuit employs positive feedback with the output fed back to the non-inverting input via the potential divider formed by $R1$ and $R2$.

Assume that C is initially uncharged and the voltage at the inverting input is slightly less than the voltage at the non-inverting input. The output voltage will rise rapidly to $+V_{CC}$ and the voltage at the inverting input will begin to rise exponentially as capacitor C charges through R. Eventually, the voltage at the inverting input will have reached a value that causes the voltage at the inverting input to exceed that present at the non-inverting input. At this point, the output voltage will rapidly fall to $-V_{CC}$. Capacitor, C, will then start to charge in the other direction and the voltage at the inverting input will begin to fall exponentially. Eventually, the voltage at the inverting input will have reached a value that causes the voltage at the inverting input to be less than that present at the non-inverting input. At this point, the output voltage will rise rapidly to $+V_{CC}$ once again and the cycle will continue indefinitely.

The upper threshold voltage (i.e. the maximum positive value for the voltage at the inverting input) will be given by:

$$V_{UT} = V_{CC} \times \frac{R2}{R1 + R2}$$

The lower threshold voltage (i.e. the maximum negative value for the voltage at the inverting input) will be given by:

$$V_{LT} = -V_{CC} \times \frac{R2}{R1 + R2}$$

Crystal controlled oscillators

A requirement of some oscillators is that they accurately maintain an exact frequency of oscillation. In such cases, a quartz crystal can be used as the frequency determining element. The quartz crystal (a thin slice of quartz in a hermetically sealed enclosure) vibrates whenever a potential difference is applied across its faces (this phenomenon is known as the **piezoelectric effect**). The frequency of oscillation is determined by the crystal's 'cut' and physical size.

Most quartz crystals can be expected to stabilize the frequency of oscillation of a circuit to within a few parts in a million. Crystals can be manufactured for operation in fundamental mode over a frequency range extending from 100 kHz to around

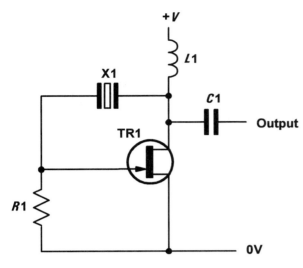

Figure 9.10   Simple crystal oscillator

20 MHz and for overtone operation from 20 MHz to well over 100 MHz. Figure 9.10 shows a simple crystal oscillator circuit in which the crystal provides feedback from the drain to the source of a junction gate FET.

Practical oscillator circuits

Figure 9.11 shows a practical sine wave oscillator based on a three-stage C–R ladder network. The circuit provides an output of approximately 1 V pk–pk at 1.97 kHz.

A practical Wien bridge oscillator is shown in Fig. 9.12. This circuit produces a sine wave output at 16 Hz. The output frequency can easily be varied by making $R1$ and $R2$ a 10 kΩ dual-gang potentiometer and connecting a fixed resistor of 680 Ω in series with each. In order to adjust the loop gain for an optimum sine wave output it may be necessary to make $R3/R4$ adjustable. One way of doing this is to replace both components with a 10 kΩ multi-turn potentiometer with the sliding contact taken to the inverting input of IC1.

An astable multivibrator is shown in Fig. 9.13. This circuit produces a square wave output of 5 V pk–pk at approximately 690 Hz.

A triangle wave generator is shown in Fig. 9.14. This circuit produces a symmetrical triangular output waveform at approximately 8 Hz. If desired, a simultaneous square wave output can be derived from the output of IC2. The circuit requires symmetrical supply voltage rails (not shown in Fig. 9.14) of between ±9 V and ±15 V.

Figure 9.15 shows a single-stage astable oscillator. This circuit produces a square wave output at approximately 13 Hz.

Finally, Fig. 9.16 shows a high-frequency crystal oscillator that produces an output of approximately 1 V pk–pk at 4 MHz. The precise frequency of operation depends upon the quartz crystal employed (the circuit will operate with fundamental mode crystals in the range 2 MHz to about 12 MHz).

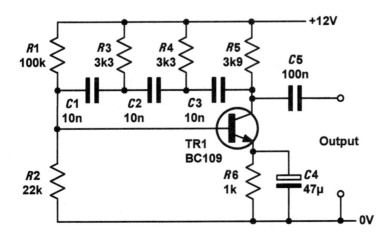

Figure 9.11   A practical sine wave oscillator based on a phase shift ladder network

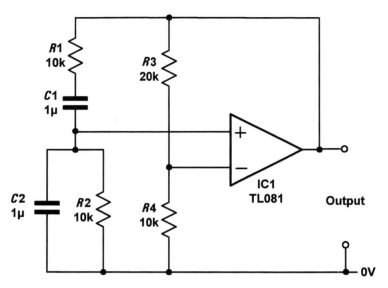

Figure 9.12   A practical sine wave oscillator based on a Wien bridge

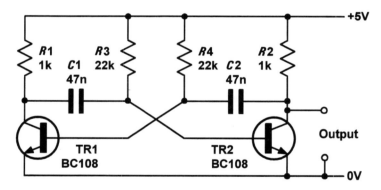

Figure 9.13   A practical square wave oscillator based on an astable multivibrator

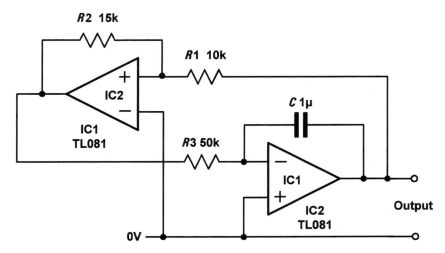

Figure 9.14   A practical triangle wave generator

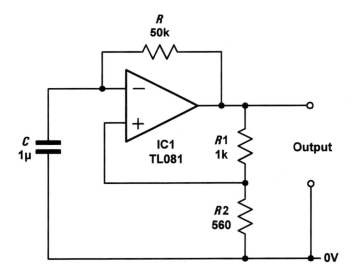

Figure 9.15   A single-stage astable oscillator which produces a square wave output

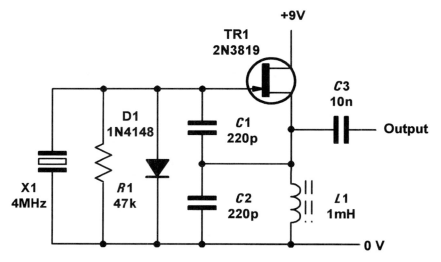

+9V

TR1
2N3819

D1
1N4148

X1
4MHz

R1
47k

C1
220p

C2
220p

C3
10n

Output

L1
1mH

0 V

Figure 9.16 A practical high-frequency crystal oscillator

Symbol introduced in this chapter

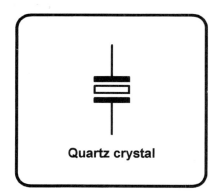

Quartz crystal

Figure 9.17

Important formulae introduced in this chapter

Gain with positive feedback applied:
(page 160)

$$A_{\text{VPFB}} = \frac{A_V}{1 - \beta A_V}$$

Loop gain:
(page 161)

$$A_{\text{VLOOP}} = \beta A_V$$

Output frequency of three-stage $C-R$ ladder network oscillator:
(page 162)

$$f = \frac{1}{2\pi\sqrt{6CR}}$$

Output frequency of Wien bridge oscillator:
(page 162)

$$f = \frac{1}{2\pi CR}$$

Time for which a multivibrator output is 'high':
(page 164)

$$T = 0.7C_T R_B$$

Periodic time for the output of a square wave multivibrator:
(page 165)

$$T = 1.4CR$$

Problems

9.1   An amplifier with a gain of 8 has 10% of its output fed back to the input. Determine the gain of the stage (a) with negative feedback, (b) with positive feedback.

9.2   A phase-shift oscillator is to operate with an output at 1 kHz. If the oscillator is based on a three-stage ladder network, determine the required values of resistance if three capacitors of 10 nF are to be used.

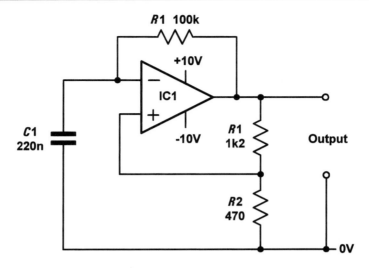

Figure 9.18

9.3 A Wien bridge oscillator is based on the circuit shown in Fig. 9.3 but $R1$ and $R2$ are replaced by a dual-gang potentiometer. If $C1 = C2 = 22$ nF determine the values of $R1$ and $R2$ required to produce an output at exactly 400 Hz.

9.4 Determine the peak–peak voltage developed across C1 in Fig. 9.18.

9.5 An astable multivibrator circuit is required to produce an asymmetrical rectangular output which has a period of 4 ms and is to be 'high' for 1 ms and 'low' for 3 ms. If the timing capacitors are both to be 100 nF, determine the values of the two timing resistors required.

(Answers to these problems can be found on page 202.)

10

Logic circuits

This chapter introduces electronic circuits and devices that are associated with digital rather than analogue circuitry. These logic circuits are used extensively in digital systems and form the basis of clocks, counters, shift registers and timers. The chapter starts by introducing the basic logic functions (AND, OR, NAND, NOR, etc.) together with the symbols and truth tables that describe the operation of the most common logic gates. We then show how these gates can be used in simple combinational logic circuits before moving on to introduce bistable devices, counters and shift registers. The chapter concludes with a brief introduction to the two principal technologies used in modern digital logic circuits, TTL and CMOS.

Logic functions

Electronic circuits can be used to make simple decisions like:

If dark then put on the light.

and

If temperature is less then 20 °C then connect the supply to the heater.

They can also be used to make more complex decisions like:

If 'hour' is greater than 11 and '24 hour clock' is not selected then display message 'pm'.

All of these logical statements are similar in form. The first two are essentially:

If {condition} then {action}.

while the third is a compound statement of the form:

If {condition 1} and not {condition 2} then {action}.

Both of these statements can be readily implemented using straightforward electronic circuitry. Because this circuitry is based on discrete states and since the behaviour of the circuits can be described by a set of logical statements, it is referred to as **digital logic**.

Switch and lamp logic

Consider the simple circuit shown in Fig. 10.1. In this circuit a battery is connected to a lamp via a switch. It should be obvious that the lamp will only operate when the switch is closed. There are two possible states for the switch, open and closed. We can summarize the operation of the circuit using Table 10.1.

Since the switch can only be in one of the two states (i.e. open or closed) at any given time, the open and closed conditions are mutually exclusive. Furthermore, since the switch cannot exist in any other state than completely open or completely closed (i.e. there is no intermediate or half-open state) the circuit uses binary or 'two-state' logic. We can represent the logical state of the switch using the binary digits, 0 and 1. We shall assume that a logical 0 is synonymous with open (or 'off') and logical 1 is synonymous with closed (or 'on'). Hence:

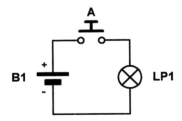

Figure 10.1   Simple switch and lamp circuit

Table 10.1   Simple switching logic

Condition	Switch	Comment
1	Open	No light produced
2	Closed	Light produced

A	LP1
0	0
1	1

Figure 10.2 Truth table for the switch and lamp

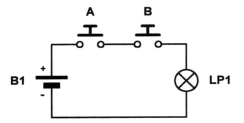

Figure 10.3 AND switch and lamp logic

Table 10.2 Simple AND switching logic

Condition	Switch A	Switch B	Comment
1	Open	Open	No light produced
2	Open	Closed	No light produced
3	Closed	Open	No light produced
4	Closed	Closed	Light produced

A	B	Y
0	0	0
0	1	1
1	0	1
1	1	1

Figure 10.4 Truth table for AND logic

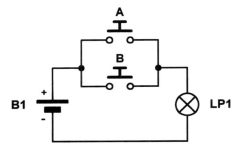

Figure 10.5 OR switch and lamp logic

Switch open (off) = 0
Switch closed (on) = 1

We can now rewrite the truth table in terms of the binary states as shown in Fig. 10.2 where:

No light (off) = 0
Light (on)   = 1

AND logic

Now consider the circuit with two switches shown in Fig. 10.3. Here the lamp will only operate when switch A is closed AND switch B is closed. However, let's look at the operation of the circuit in a little more detail. Since there are two switches (A and B) and there are two possible states for each switch (open or closed), there is a total of four possible conditions for the circuit. We can summarize these conditions in Table 10.2.

Since each switch can only be in one of the two states (i.e. open or closed) at any given time, the open and closed conditions are mutually exclusive. Furthermore, since the switches cannot exist in any other state than completely open or completely closed (i.e. there are no intermediate states) the circuit uses 'binary logic'. We can thus represent the logical states of the two switches by the binary digits, 0 and 1.

Once again, if we adopt the obvious convention that an open switch can be represented by 0 and a closed switch by 1, we can rewrite the truth table in terms of the binary states shown in Fig. 10.4 where

No light (off) = 0
Light (on)   = 1

OR logic

Figure 10.5 shows another circuit with two switches. This circuit differs from that shown in Fig. 10.3 by virtue of the fact that the two switches are connected in parallel rather than in series. In this case the lamp will operate when either of the two switches is closed. As before, there is a total of four possible conditions for the circuit. We can summarize these conditions in Table 10.3.

Once again, adopting the convention that an open switch can be represented by 0 and a closed switch by 1, we can rewrite the truth table in terms of the binary states as shown in Fig. 10.6.

Table 10.3 Simple OR switching logic

Condition	Switch A	Switch B	Comment
1	Open	Open	No light produced
2	Open	Closed	Light produced
3	Closed	Open	Light produced
4	Closed	Closed	Light produced

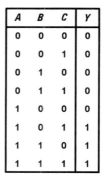

A	B	C	Y
0	0	0	0
0	0	1	0
0	1	0	0
0	1	1	0
1	0	0	0
1	0	1	1
1	1	0	1
1	1	1	1

Figure 10.8

A	B	Y
0	0	0
0	1	1
1	0	1
1	1	1

Figure 10.6 Truth table for OR logic

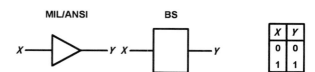

X	Y
0	0
1	1

Figure 10.9 Symbols and truth table for a buffer

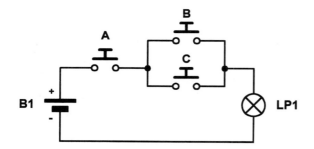

Figure 10.7

Example 10.1

Figure 10.7 shows a simple switching circuit. Describe the logical state of switches A, B, and C in order to operate the lamp. Illustrate your answer with a truth table.

Solution

In order to operate the lamp, switch A AND either switch B OR switch C must be operated. The truth table is shown in Fig. 10.8.

Logic gates

Logic gates are circuits designed to produce the basic logic functions, AND, OR, etc. These circuits are designed to be interconnected into larger, more complex, logic circuit arrangements. Since these circuits form the basic building blocks of all digital systems, we have summarized the action of each of the gates in the next section. For each gate we have included its British Standard (BS) symbol together with its American Standard (MIL/ANSI) symbol. We have also included the truth tables and Boolean expressions (using '+' to denote OR, '.' to denote AND, and '−' to denote NOT). Note that, while inverters and buffers each have only one input, exclusive-OR gates have two inputs and the other basic gates (AND, OR, NAND and NOR) are commonly available with up to eight inputs.

Buffers (Fig. 10.9)

Buffers do not affect the logical state of a digital signal (i.e. a logic 1 input results in a logic 1 output whereas a logic 0 input results in a logic 0 output). Buffers are normally used to provide extra current drive at the output but can also be used to regularize the logic levels present at an interface.

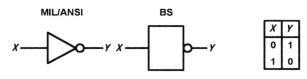

Figure 10.10   Symbols and truth table for an inverter

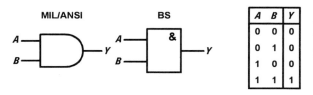

Figure 10.11   Symbols and truth table for an AND gate

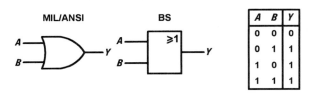

Figure 10.12   Symbols and truth table for an OR gate

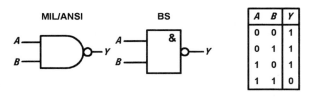

Figure 10.13   Symbols and truth table for a NAND gate

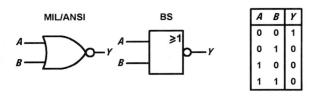

Figure 10.14   Symbols and truth table for a NOR gate

The Boolean expression for the output, Y, of a buffer with an input, X, is:

$Y = X$

Inverters (Fig. 10.10)

Inverters are used to complement the logical state (i.e. a logic 1 input results in a logic 0 output and vice versa). Inverters also provide extra current drive and, like buffers, are used in interfacing applications where they provide a means of regularizing logic levels present at the input or output of a digital system. The Boolean expression for the output, Y, of a buffer with an input, X, is:

$Y = \overline{X}$

AND gates (Fig. 10.11)

AND gates will only produce a logic 1 output when all inputs are simultaneously at logic 1. Any other input combination results in a logic 0 output. The Boolean expression for the output, Y, of an AND gate with inputs, A and B, is:

$Y = A.B$

OR gates (Fig. 10.12)

OR gates will produce a logic 1 output whenever any one, or more, inputs are at logic 1. Putting this another way, an OR gate will only produce a logic 0 output whenever all of its inputs are simultaneously at logic 0. The Boolean expression for the output, Y, of an OR gate with inputs, A and B, is:

$Y = A + B$

NAND gates (Fig. 10.13)

NAND (i.e. NOT-AND) gates will only produce a logic 0 output when all inputs are simultaneously at logic 1. Any other input combination will produce a logic 1 output. A NAND gate, therefore, is nothing more than an AND gate with its output inverted! The circle shown at the output denotes this inversion. The Boolean expression for the output, Y, of a NAND gate with inputs, A and B, is:

$Y = \overline{A}.\overline{B}$

NOR gates (Fig. 10.14)

NOR (i.e. NOT-OR) gates will only produce a logic 1 output when all inputs are simultaneously at logic

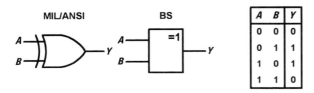

A	B	Y
0	0	0
0	1	1
1	0	1
1	1	0

Figure 10.15 Symbols and truth table for an exclusive-OR gate

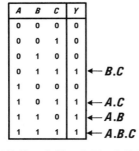

$Y = (B.C) + (A.C) + (A.B) + A.B.C$

Figure 10.16

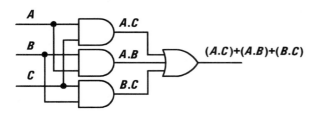

Figure 10.17

0. Any other input combination will produce a logic 0 output. A NOR gate, therefore, is simply an OR gate with its output inverted. A circle is again used to indicate inversion. The Boolean expression for the output, Y, of a NOR gate with inputs, A and B, is:

$Y = \overline{A + B}$

Exclusive-OR gates (Fig. 10.15)

Exclusive-OR gates will produce a logic 1 output whenever either one of the inputs is at logic 1 and

the other is at logic 0. Exclusive-OR gates produce a logic 0 output whenever both inputs have the same logical state (i.e. when both are at logic 0 or both are at logic 1). The Boolean expression for the output, Y, of an exclusive-OR gate with inputs, A and B, is:

$Y = A.\overline{B} + B.\overline{A}$

Combinational logic

By using a standard range of logic levels (i.e. voltage levels used to represent the logic 1 and logic 0 states) logic circuits can be combined together in order to solve complex logic functions.

Example 10.2

A logic circuit is to be constructed that will produce a logic 1 output whenever two, or more, of its three inputs are at logic 1.

Solution

This circuit could be more aptly referred to as a **majority vote** circuit. Its truth table is shown in Fig. 10.16. Figure 10.17 shows the logic circuitry required.

Example 10.3

Show how an arrangement of basic logic gates (AND, OR and NOT) can be used to produce the exclusive-OR function.

Solution

In order to solve this problem, consider the Boolean expression for the exclusive-OR function:

$Y = A.\overline{B} + B.\overline{A}$

This expression takes the form:

$Y = P + Q$ where $P = A.\overline{B}$ and $Q = B.\overline{A}$

$A.\overline{B}$ and $B.\overline{A}$ can be obtained using two two-input AND gates and the result (i.e. P and Q) can then be applied to a two-input OR gate.
$\overline{A}$ and $\overline{B}$ can be produced using inverters.
The complete solution is shown in Fig. 10.18.

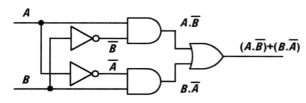

Figure 10.18

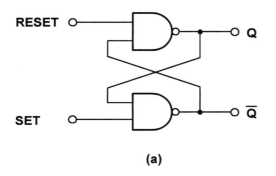

(a)

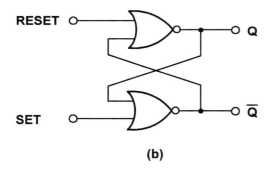

(b)

Figure 10.19 R-S bistables using cross-coupled NAND and NOR gates

Bistables

The output of a bistable has two stables states (logic 0 or logic 1) and, once **set** in one or other of these states, the device will remain at a particular logic level for an indefinite period until **reset**. A bistable thus constitutes a simple form of 'memory cell'; as it will remain in its **latched** state (whether set or reset) until a signal is applied to it in order to change its state (or until the supply is disconnected).

R-S bistables

The simplest form of bistable is the R-S bistable. This device has two inputs, SET and RESET, and complementary outputs, Q and $\overline{Q}$. A logic 1 applied to the SET input will cause the Q output to become (or remain at) logic 1 while a logic 1 applied to the RESET input will cause the Q output to become (or remain at) logic 0. In either case, the bistable will remain in its SET or RESET state until an input is applied in such a sense as to change the state.

Two simple forms of R-S bistable based on cross-coupled logic gates are shown in Fig. 10.19. Figure 10.19(a) is based on NAND gates while Fig. 10.19(b) is based on NOR gates.

The simple cross-coupled logic gate bistable has a number of serious shortcomings (consider what would happen if a logic 1 was simultaneously present on both the SET and RESET inputs!) and practical forms of bistable make use of much improved purpose-designed logic circuits such as D-type and J-K bistables.

D-type bistables

The D-type bistable has two inputs: D (standing variously for data or delay) and CLOCK (CLK). The data input (logic 0 or logic 1) is clocked into the bistable such that the output state only changes when the clock changes state. Operation is thus said to be synchronous. Additional subsidiary inputs (which are invariably active low) are provided which can be used to directly set or reset the bistable. These are usually called PRESET (PR) and CLEAR (CLR). D-type bistables are used both as latches (a simple form of memory) and as binary dividers. The simple circuit arrangement in Fig. 10.20 together with the timing diagram shown in Fig. 10.21 illustrate the operation of D-type bistables.

J-K bistables

J-K bistables have two clocked inputs (J and K), two direct inputs (PRESET and CLEAR), a CLOCK (CK) input, and outputs (Q and $\overline{Q}$). As with R-S bistables, the two outputs are complementary (i.e. when one is 0 the other is 1, and vice versa). Similarly, the PRESET and CLEAR inputs are invariably both active low (i.e. a 0 on the PRESET input will set the Q output to 1 whereas a 0 on the CLEAR

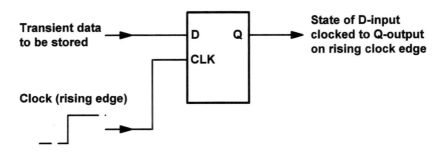

Figure 10.20 D-type bistable operation

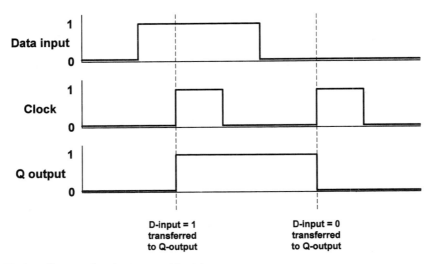

Figure 10.21 Timing diagram for the D-type bistable

input will set the Q output to 0). Table 10.4 summarizes the operation of a J-K bistable for various input states.

J-K bistables are the most sophisticated and flexible of the bistable types and they can be configured in various ways including binary dividers, shift registers, and latches. Figure 10.22 shows the arrangement of a four-stage binary counter based on J-K bistables. The **timing diagram** for this circuit is shown in Fig. 10.23. Each stage successively divides the clock input signal by a factor of two. Note that a logic 1 input is transferred to the respective Q-output on the falling edge of the clock pulse and all J and K inputs must be taken to logic 1 to enable binary counting.

Figure 10.24 shows the arrangement of a four-stage shift register based on J-K bistables. The timing diagram for this circuit is shown in Fig. 10.25.

Note that each stage successively feeds data (Q output) to the next stage. Note that all data transfer occurs on the falling edge of the clock pulse.

Example 10.4

A certain logic arrangement is to produce the pulse train shown in Fig. 10.26. Devise a logic circuit arrangement that will generate this pulse train from a regular square wave input.

Solution

A two-stage binary divider (based on J-K bistables) can be used together with a two-input AND gate as shown in Fig. 10.27. The waveforms for this logic arrangement are shown in Fig. 10.28.

Table 10.4 Input and output states for a J-K bistable

| Inputs | | Output | Comments |
J	K	Q_{N+1}	
0	0	Q_N	No change in state of the Q output on the next clock transition
0	1	0	Q output changes to 0 (i.e. Q is reset) on the next clock transition
1	0	1	Q output changes to 1 (i.e. Q is set) on the next clock transition
1	1	Q_N	Q output changes to the opposite state on the next clock transition

Note: Q_{N+1} means 'Q after next clock transition' while Q_N means 'Q in whatever state it was before'.

| Inputs | | Output | Comments |
Preset	Clear	Q	
0	0	?	Indeterminate
0	1	0	Q output changes to 1 (i.e. Q is reset) regardless of the clock state
1	0	1	Q output changes to 1 (i.e. Q is set) regardless of the clock state
1	1	–	Enables clocked operation (refer to previous table for state of Q_{N+1})

Note: The preset and clear inputs operate regardless of the clock.

Integrated circuit logic devices

The task of realizing a complex logic circuit is made simple with the aid of digital integrated circuits. Such devices are classified according to the semiconductor technology used in their fabrication (the **logic family** to which a device belongs is largely instrumental in determining its operational characteristics, such as power consumption, speed, and immunity to noise).

The two basic logic families are CMOS (complementary metal oxide semiconductor) and TTL (transistor transistor logic). Each of these families is then further sub-divided. Representative circuits for a two-input AND gate in both technologies are shown in Figs 10.29 and 10.30.

The most common family of TTL logic devices is known as the 74-series. Devices from this family are coded with the prefix number 74. Variants within the family are identified by letters which follow the initial 74 prefix, as shown in Table 10.5. The most common family of CMOS devices is known as the 4000-series. Variants within the family are identified by the suffix letters given in Table 10.6

Example 10.5

Identify each of the following integrated circuits:

(i) 4001UBE;
(ii) 74LS14.

Solution

Integrated circuit (i) is an improved (unbuffered) version of the CMOS 4001 device.

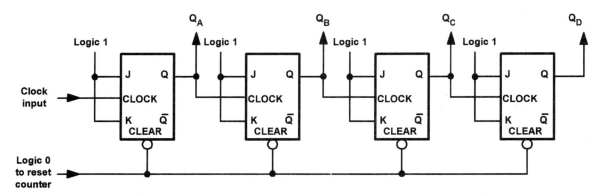

Figure 10.22 Four-stage binary counter using J-K bistables

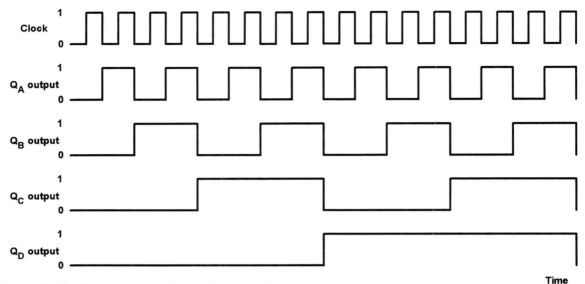

Figure 10.23 Timing diagram for the four-stage binary counter

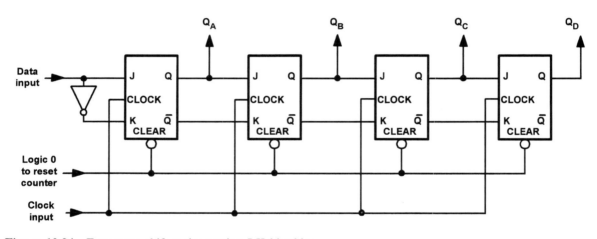

Figure 10.24 Four-stage shift register using J-K bistables

Integrated circuit (ii) is a low-power Schottky version of the TTL 7414 device.

Date codes

It is also worth noting that the vast majority of logic devices and other digital integrated circuits are marked with a four digit date code. The first two digits of this code give the last two digits of the year of manufacture while the last two digits

specify the week of manufacture. The code often appears alongside or below the device code.

Example 10.6

An integrated circuit marked '4050B 8832'. What type of device is it and when was it manufactured?

Solution

The device is a buffered CMOS 4050 manufactured in the 32nd week of 1988.

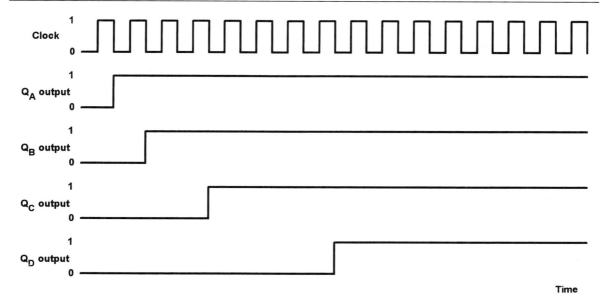

Figure 10.25 Timing diagram for the four-stage shift register

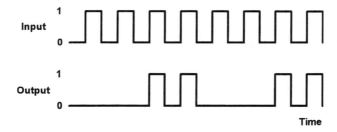

Figure 10.26

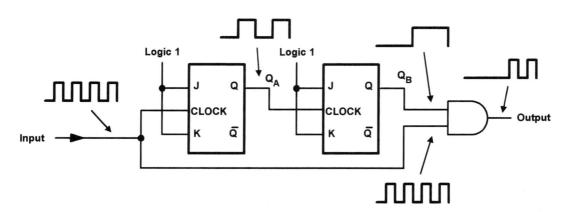

Figure 10.27

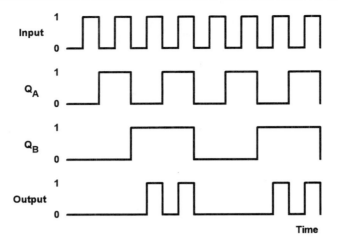

Figure 10.28

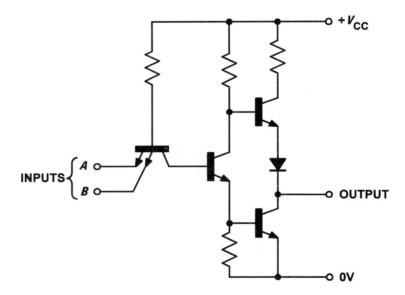

Figure 10.29   Two-input TTL NAND gate

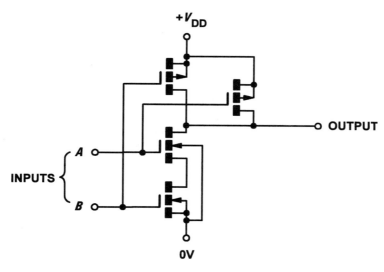

Figure 10.30 Two-input CMOS NAND gate

Table 10.5 TTL device coding

Infix	Meaning
None	Standard TTL device
ALS	Advanced low-power Schottky
C	CMOS version of a TTL device
F	'Fast' – a high-speed version of the device
H	High-speed version
S	Schottky input configuration (improved speed and noise immunity)
HC	High-speed CMOS version (CMOS compatible inputs)
HCT	High-speed CMOS version (TTL compatible inputs)
LS	Low-power Schottky

Table 10.6 CMOS device coding
The most common family of CMOS devices is known as the 4000-series. Variants within the family are identified by suffix letters as follows:

Suffix	Meaning
None	Standard CMOS device
A	Standard (unbuffered) CMOS device
B, BE	Improved (buffered) CMOS device
UB, UBE	Improved (unbuffered) CMOS device

Table 10.7 CMOS and TTL characteristics

Condition	CMOS	TTL
Logic 1	More than $(2/3)\ V_{DD}$	More than 2 V
Logic 0	Less than $(1/3)\ V_{DD}$	Less than 0.8 V
Indeterminate	Between $(1/3)\ V_{DD}$ and $(2/3)\ V_{DD}$	Between 0.8 V and 2 V

Note: V_{DD} is the positive supply associated with CMOS devices.

Logic levels

Logic levels are simply the range of voltages used to represent the logic states 0 and 1. The logic levels for CMOS differ markedly from those associated with TTL. In particular, CMOS logic levels are relative to the supply voltage used while the logic levels associated with TTL devices tend to be absolute (see Table 10.7).

Noise margin

The noise margin of a logic device is a measure of its ability to reject noise; the larger the noise margin

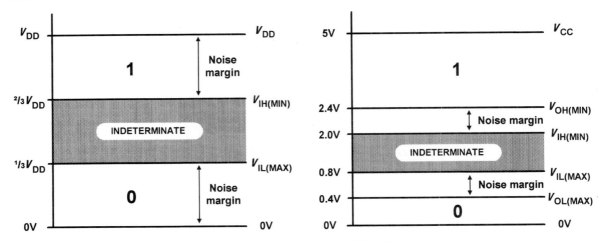

Figure 10.31   Logic levels and noise margins for TTL and CMOS devices

Table 10.8   Characteristics of common logic families

Characteristic	Logic family			
	74	74LS	74HC	40BE
Maximum supply voltage	5.25 V	5.25 V	5.5 V	18 V
Minimum supply voltage	4.75 V	4.75 V	4.5 V	3 V
Static power dissipation (mW per gate at 100 kHz)	10	2	negligible	negligible
Dynamic power dissipation (mW per gate at 100 kHz)	10	2	0.2	0.1
Typical propagation delay (ns)	10	10	10	105
Maximum clock frequency (MHz)	35	40	40	12
Speed–power product (pJ at 100 kHz)	100	20	1.2	11
Minimum output current (mA at $V_o = 0.4$ V)	16	8	4	1.6
Fan-out (number of LS loads that can be driven)	40	20	10	4
Maximum input current (mA at $V_i = 0.4$ V)	−1.6	−0.4	0.001	−0.001

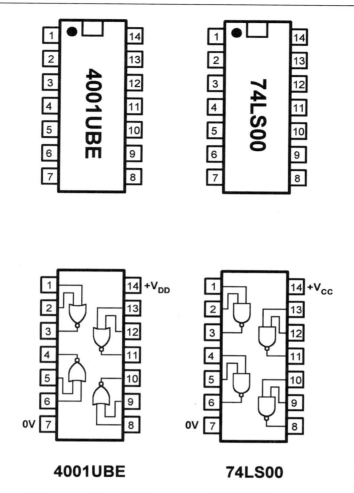

Figure 10.32 Packages and pin connections for two common logic devices

the better is its ability to perform in an environment in which noise is present. Noise margin is defined as the difference between the minimum values of high state output and high state input voltage and the maximum values of low state output and low state input voltage. Hence:

noise margin $= V_{oh(MIN)} - V_{ih(MIN)}$

or

noise margin $= V_{ol(MAX)} - V_{il(MAX)}$

where $V_{oh(MIN)}$ is the minimum value of high state (logic 1) output voltage, $V_{ih(MIN)}$ is the minimum

value of high state (logic 1) input voltage, $V_{ol(MAX)}$ is the maximum value of low state (logic 0) output voltage, and $V_{il(MIN)}$ is the minimum value of low state (logic 0) input voltage.

The noise margin for standard 7400 series TTL is typically 400 mV while that for CMOS is (1/3) V_{DD}, as shown in Fig. 10.31.

Table 10.8 compares the more important characteristics of common members of the TTL family with their buffered CMOS logic counterparts. Finally, Fig. 10.32 shows the packages and pin connections for two common logic devices, the 74LS00 (quad two-input NAND gate) and the 4001UBE (quad two-input NOR gate).

Circuit symbols introduced in this chapter

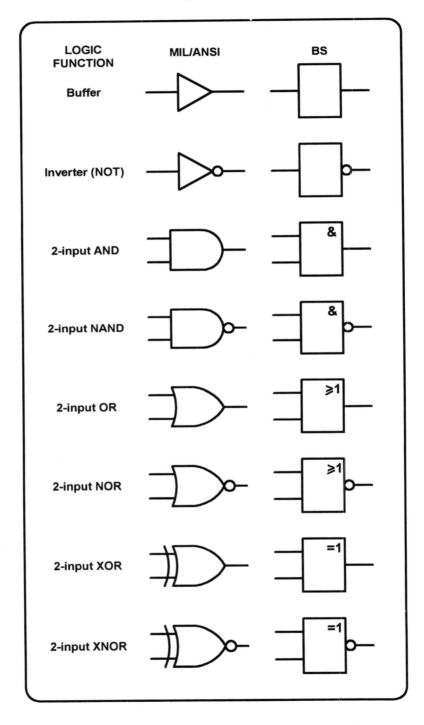

Figure 10.33

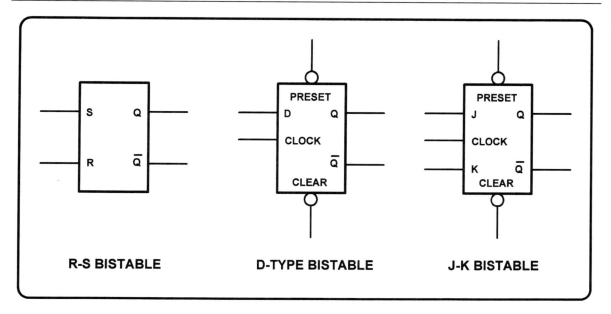

Figure 10.34   Symbols for R-S, D-type and J-K bistables

Formulae introduced in this chapter

Noise margin:
(page 183)

Noise margin $= V_{oh(MIN)} - V_{ih(MIN)}$

Noise margin $= V_{ol(MAX)} - V_{il(MAX)}$

Problems

10.1   Show how a four-input AND gate can be made from three two-input AND gates.

10.2   Show how a four-input OR gate can be made from three two-input OR gates.

10.3   Construct the truth table for the logic gate arrangement shown in Fig. 10.35.

10.4   Using only two-input NAND gates, show how each of the following logical functions can be satisfied:

(a) two-input AND;
(b) two-input OR;
(c) four-input AND.

In each case, use the minimum number of gates.
(Hint: a two-input NAND gate can be made into an inverter by connecting its two inputs together)

10.5   The rocket motor of a missile will operate if, and only if, the following conditions are satisfied:

(i) 'launch' signal is at logic 1;
(ii) 'unsafe height' signal is at logic 0;
(iii) 'target lock' signal is at logic 1.

Devise a suitable logic arrangement that will satisfy this requirement. Use the minimum number of logic gates.

10.6   An automatic sheet metal guillotine will operate if the following conditions are satisfied:

(i) 'guard lowered' signal is at logic 1;
(ii) 'feed jam' signal is at logic 0;
(iii) 'manual start' signal is at logic 1.

The sheet metal guillotine will also operate if the following conditions are satisfied:

(i) 'manual start' signal is at logic 1;
(ii) 'test key' signal is at logic 1.

Devise a suitable logic arrangement that will satisfy this requirement. Use the minimum number of logic gates.

10.7   Devise a logic arrangement using no more than four two-input gates that will satisfy the truth table shown in Fig. 10.36.

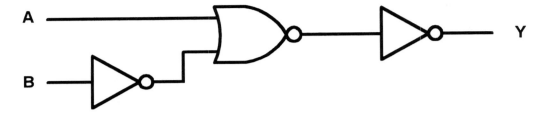

Figure 10.35

A	B	C	Y
0	0	0	0
0	0	1	1
0	1	0	0
0	1	1	1
1	0	0	0
1	0	1	1
1	1	0	1
1	1	1	1

Figure 10.36

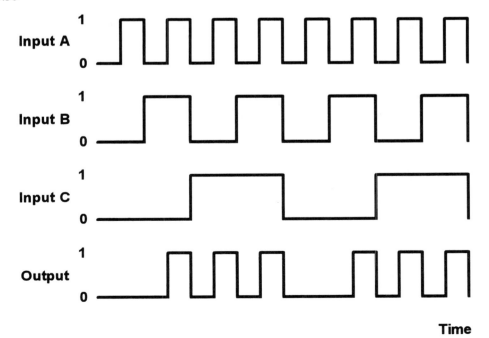

Figure 10.37

10.8   Devise a logic arrangement that will produce the output waveform from the three input waveforms shown in Fig. 10.37.

10.9   A logic device is marked '74LS90 2789'. To which family and sub-family of logic does it belong? When was the device manufactured?

10.10   A logic family recognizes a logic 1 input as being associated with any voltage between 2.0 V and 5.5 V. The same family produces an output in the range 2.6 V to 5.0 V corresponding to a logic 1 output. Determine the noise margin.

(Answers to these problems can be found on page 202.)

Appendix 1

Student assignments

The twelve student assignments provided here have been designed to satisfy a selection of the published GNVQ Evidence Indicators for the following GNVQ units:

- Electronics (Intermediate)
- Electronic Principles and Applications (Intermediate)
- Electrical Principles and Materials (Intermediate)
- Electronics (Advanced)
- Electrical and Electronic Principles (Advanced)
- Applied Electronics (Advanced)
- Microelectronics (Advanced)

Please note that the assignments are not exhaustive and may need modification to meet an individual awarding body's requirements and locally available resources. The first six assignments satisfy the requirements for the intermediate level while the remaining six are designed to meet the requirements of the advanced level. Assignments can normally be carried out in three to five hours, including analysis and report writing.

Intermediate level

Assignment 1   Electronic circuit construction

For each one of five simple electronic circuits shown in Figs A1.1 to A1.5:

(a) Identify and select the components required to build the circuit.
(b) Identify an appropriate method of construction selected from the following list:

- prototype board;
- tag board;
- strip board;
- printed circuit board;
- wire wrapping.

Assemble and test each circuit according to its circuit diagram. Note that a different construction method must be selected for each circuit.

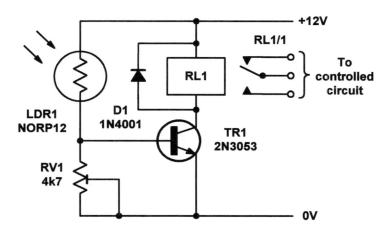

Figure A1.1   Light operated switch circuit for Assignments 1 and 2

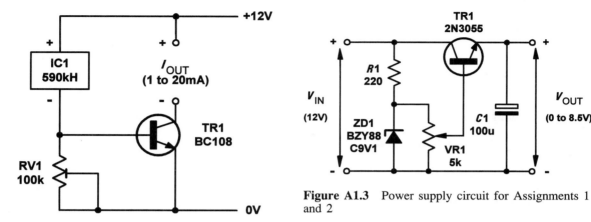

Figure A1.2   Heat-sensing unit circuit for Assignments 1 and 2

Figure A1.3   Power supply circuit for Assignments 1 and 2

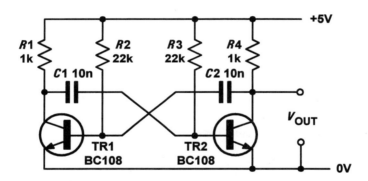

Figure A1.4   Astable multivibrator circuit for Assignments 1 and 2

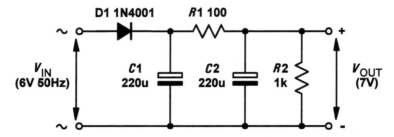

Figure A1.5   Half-wave rectifier circuit for Assignments 1 and 2

Assignment 2   Electronic circuit testing

For each of the circuits in Figs A1.1 to A1.5, describe the type and nature of the input and output signals (as appropriate). For each circuit select and use appropriate measuring instruments (e.g. multimeter and oscilloscope) to test each circuit. Write a report to summarize your findings.

Assignment 3   Semiconductor investigation

Prepare a report describing the construction of (a) a junction diode and (b) a bipolar transistor. Describe, in your own words, the principle of operation of each device. With the aid of a simple circuit diagram, describe a typical application for each type of semiconductor.

Assignment 4   Basic logic functions

Write a report identifying the types and symbols (both BS3939 and MIL/ANSI) used for all basic logic gates (AND, OR, NOT, NAND and NOR). Include in your report a description of the operation of each logic gate together with a truth table.

Assignment 5   Applications of logic circuits

With the aid of labelled diagrams, describe TWO applications of logic gates. One application should be based on combinational logic while the other should use sequential logic.

Assignment 6   Electronic measuring instruments

Write a report describing the operation and use of (a) a multimeter and (b) an oscilloscope. Illustrate your report with records of measurements carried out on three common electronic components and two simple electronic circuits.

Advanced level assignments

Assignment 7   Power supply investigation

With the aid of an electrical specification and operating manual for a typical low-voltage d.c. power supply, write a report explaining the characteristics of the unit. Also explain the meaning of each of the unit's specifications. Carry out a simple load test on the supply, plot a graph to illustrate your results and comment on your findings.

Assignment 8   Amplifier circuit investigation

Write a report that describes and explains one small-signal Class-A discrete amplifier circuit, one Class-B power amplifier circuit, and one amplifier circuit based on an operational amplifier. The report should identify and give typical specifications for each type of amplifier. Carry out a simple gain and frequency response test on one of the amplifier circuits, plot a graph to illustrate your results and comment on your findings.

Assignment 9   Small-signal amplifier design and construction

Design, construct and test a single-stage transistor amplifier. Write a report describing the model used for the small-signal amplifier and include a detailed comparison of the predicted characteristics compared with the measured performance of the stage.

Assignment 10   Electronic counter investigation

Prepare a report describing the characteristics of J-K bistable elements when compared with D-type and R-S bistables. Using the data obtained, design, construct and test a four-stage binary counter. Modify the design using a standard logic gate to produce a decade counter. Include in your report timing diagrams for both the binary and decade counters.

Assignment 11   Waveform generator investigation

Prepare a report describing one oscillator circuit, one bistable circuit and one monostable circuit. The report should include a description of one industrial application for each of the circuits together with sample calculations of frequency and pulse rate for the two oscillator circuits, and pulse width for the monostable circuit.

Assignment 12   Performance testing

Carry out performance tests on two different amplifier circuits, two waveform generators and two digital circuits. Compare the measured performance of each circuit with the manufacturer's specification and present your findings in a written report. The report should include details of the calibration and operation of each test instrument in accordance with the manufacturer's handbooks as well as evidence of the adoption of safe working practice.

Appendix 2

Revision problems

These 50 problems provide you with a means of checking your understanding prior to an end-of-course assessment or formal examination. If you have difficulty with any of the questions you should refer to the page numbers indicated.

1. A 120 kΩ resistor is connected to a 6 V battery. Determine the current flowing. [Page 6]
2. A current of 45 mA flows in a resistor of 2.7 kΩ. Determine the voltage dropped across the resistor. [Page 6]
3. A 24 V d.c. supply delivers a current of 1.5 A. Determine the power supplied. [Page 6]
4. A 27 Ω resistor is rated at 3 W. Determine the maximum current that can safely be applied to the resistor. [Page 8]
5. A load resistor is required to dissipate a power of 50 W from a 12 V supply. Determine the value of resistance required. [Page 8]
6. An electrical conductor has a resistance of 0.05 Ω per metre. Determine the power wasted in a 175 m length of this conductor when a current of 8 A is flowing in it. [Page 7]
7. Figure A2.1 shows a node in a circuit. Determine the value of I_X. [Page 48]
8. Figure A2.2 shows part of a circuit. Determine the value of V_X. [Page 48]
9. A capacitor of 200 µF is charged to a potential of 50 V. Determine the amount of charge stored. [Page 33]
10. A sinusoidal a.c. supply has a frequency of 400 Hz and an r.m.s. value of 120 V. Determine the periodic time and peak value of the supply. [Page 70]
11. Four complete cycles of a waveform occur in a time interval of 20 ms. Determine the frequency of the waveform. [Page 70]
12. Determine the periodic time, frequency and amplitude of each of the waveforms shown in Fig. A2.3. [Page 70]
13. Determine the effective resistance of each circuit shown in Fig. A2.4. [Page 23]
14. Determine the effective capacitance of each circuit shown in Fig. A2.5. [Page 35]
15. Determine the effective inductance of each circuit shown in Fig. A2.6. [Page 42]
16. A quantity of 100 nF capacitors is available, each rated at 100 V working. Determine how several of these capacitors can be connected to produce an equivalent capacitance of: (a) 50 nF rated at 200 V; (b) 250 nF rated at 100 V; and (c) 300 nF rated at 100 V. [Page 35]
17. Two 60 mH inductors and two 5 mH inductors are available, each rated at 1 A. Determine how some or all of these can be connected to produce an equivalent inductance of: (a) 30 mH rated at 2 A; (b) 40 mH rated at 1 A; and (c) 125 mH rated at 1 A. [Page 42]
18. Determine the resistance looking into the network shown in Fig. A2.7, (a) with C and D open-circuit and (b) with C and D shorted together. [Page 23]

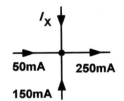

Figure A2.1

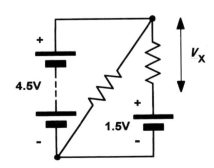

Figure A2.2

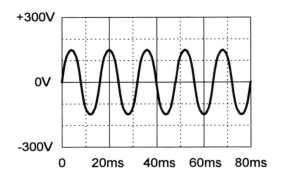

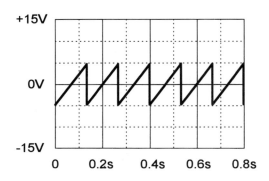

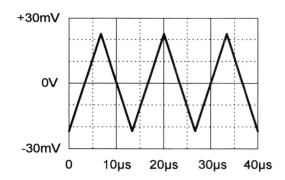

Figure A2.3

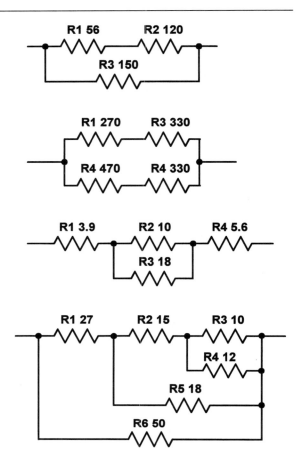

Figure A2.4

19. Determine the current flowing in each resistor and voltage dropped across each resistor in Fig. A2.8. [Page 48]
20. Determine the current flowing in the voltmeter movement show in Fig. A2.9. [Page 48]
21. Assuming that the capacitor shown in Fig. A2.10 is initially fully discharged (by switching to position B), determine the current in $R1$ at the instant that S1 is switched to position A. Also determine the capacitor voltage 1 minute after operating the switch. [Page 55]
22. Determine the time taken for the output voltage in Fig. A2.11 to reach 4 V after the arrival of the pulse shown (assume that the capacitor is initially uncharged). [Page 55]
23. In Fig. A2.12, determine the current supplied to the inductor 100 ms after pressing the 'start' button. [Page 61]
24. Determine the reactance at 2 kHz of (a) a

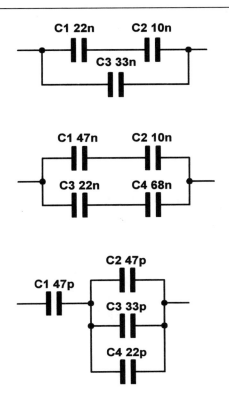

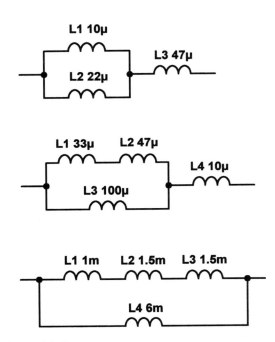

Figure A2.5

Figure A2.6

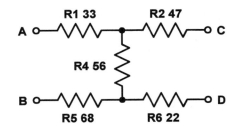

Figure A2.7

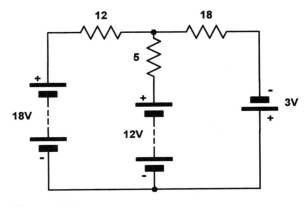

Figure A2.8

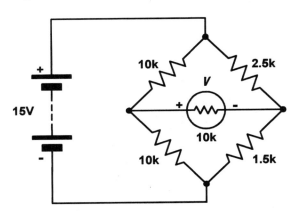

Figure A2.9

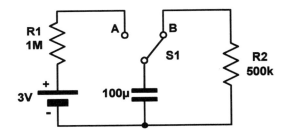

Figure A2.10

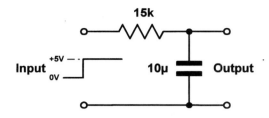

Figure A2.11

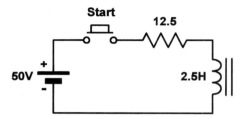

Figure A2.12

Forward current (mA)

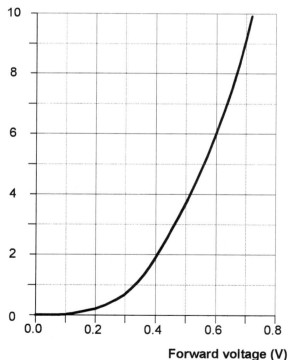

Forward voltage (V)

Figure A2.13

60 mH inductor and (b) a 47 nF capacitor. [Page 71]

25. A 50 µF capacitor is connected to a 12 V, 50 Hz a.c. supply. Determine the current flowing. [Page 71]

26. An inductor of 2 H is connected to a 12 V, 50 Hz a.c. supply. If the inductor has a winding resistance of 40 Ω, determine the current flowing and the phase angle between the supply voltage and supply current. [Page 73]

27. An inductor of 100 µH is connected in series with a variable capacitor. If the capacitor is variable over the range 50 pF to 500 pF, determine the maximum and minimum values of resonant frequency for the circuit. [Page 76]

28. An audio amplifier delivers an output power of 40 W r.m.s. to an 8 Ω resistive load. What r.m.s. voltage will appear across the load? [Page 70]

29. A transformer has 400 primary turns and 60 secondary turns. The primary is connected to a 220 V a.c. supply and the secondary is connected to a load resistance of 20 Ω. Assuming that the transformer is perfect, determine: (a) the secondary voltage; (b) the secondary current; and (c) the primary current. [Page 77]

30. Figure A2.13 shows the characteristic of a diode. Determine the resistance of the diode when (a) $V_F = 2$ V and (b) $I_F = 9$ mA. [Page 87]

31. A transistor operates with a collector current of 25 mA and a base current of 200 µA. Determine: (a) the value of emitter current; (b) the value of common-emitter current gain; and (c) the new value of collector current if the base current increases by 50%. [Page 96]

32. A zener diode rated at 5.6 V is connected to a 12 V d.c. supply via a fixed series resistor of 56 Ω. Determine the current flowing in the resistor, the power dissipated in the resistor and the power dissipated in the zener diode. [Page 116]

33. An amplifier has identical input and output resistances and provides a voltage gain of 26 dB. Determine the output voltage produced if an input of 50 mV is applied. [Page 121]

34. Figure A2.14 shows the frequency response of an amplifier. Determine the mid-band voltage gain and the upper and lower cut-off frequencies. [Page 125]

35. Figure A2.15 shows the frequency response of an amplifier. Determine the bandwidth of the amplifier. [Page 127]

36. The transfer characteristic of a transistor is shown in Fig. A2.16. Determine (a) the static

Voltage gain

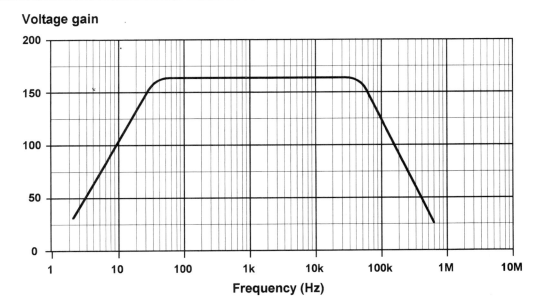

Figure A2.14

Voltage gain

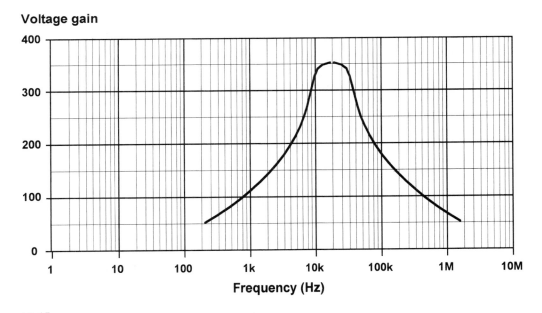

Figure A2.15

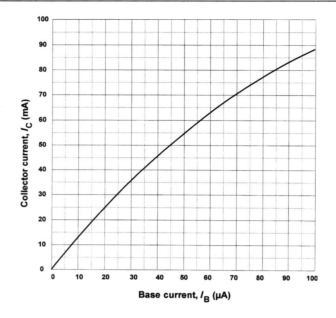

Figure A2.16

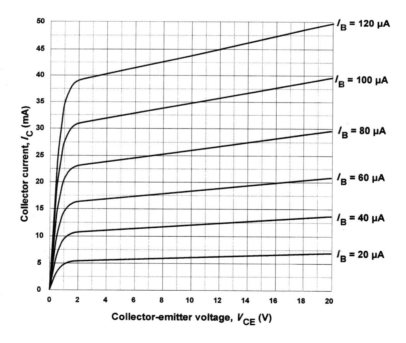

Figure A2.17

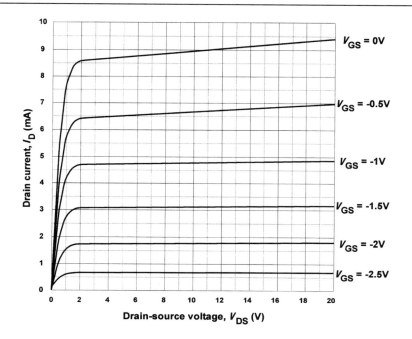

Figure A2.18

value of common-emitter current gain at $I_C = 50$ mA and (b) the dynamic (small-signal) value of common-emitter current gain at $I_C = 50$ mA. [Page 132]

37. The output characteristics of a bipolar transistor are shown in Fig. A2.17. If the transistor operates with $V_{CC} = 15$ V, $R_L = 500$ Ω and $I_B = 40$ μA determine:

   (a) the quiescent value of collector–emitter voltage;
   (b) the quiescent value of collector current;
   (c) the peak–peak output voltage produced by a base input current of 40 μA. [Page 137]

38. The output characteristics of a field effect transistor are shown in Fig. A2.18. If the transistor operates with $V_{DD} = 18$ V, $R_L = 3$ kΩ and $V_{GS} = -1.5$ V determine:

   (a) the quiescent value of drain–source voltage;
   (b) the quiescent value of drain current;
   (c) the peak–peak output voltage produced by a gate input voltage of 1 V pk–pk;
   (d) the voltage gain of the stage. [Page 137]

39. Figure A2.19 shows the circuit of a common-emitter amplifier stage. Determine the values of I_B, I_C, I_E and the voltage at the emitter. [Page 137]

40. A transistor having $h_{ie} = 2.5$ kΩ and $h_{fe} = 220$ is used in a common-emitter amplifier stage with $R_L = 3.3$ kΩ. Assuming that h_{oe} and h_{re} are negligible, determine the voltage gain of the stage. [Page 135]

41. An astable multivibrator is based on coupling capacitors $C1 = C2 = 10$ nF and timing resistors $R1 = 10$ kΩ and $R2 = 4$ kΩ. Determine the frequency of the output signal. [Page 163]

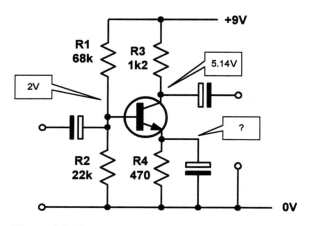

Figure A2.19

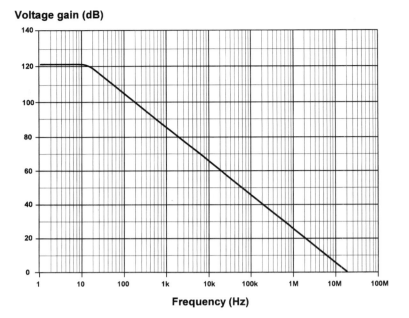

Figure A2.20

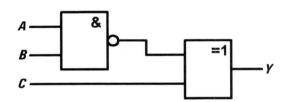

Figure A2.21

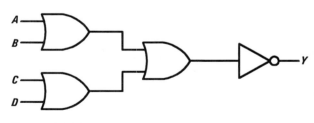

Figure A2.23

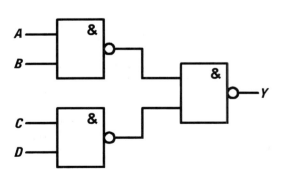

Figure A2.22

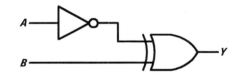

Figure A2.24

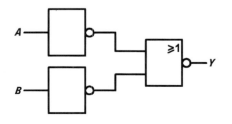

Figure A2.25

A	B	Y
0	0	1
0	1	1
1	0	0
1	1	1

A	B	C	Y
0	0	0	0
0	0	1	0
0	1	0	0
0	1	1	1
1	0	0	0
1	0	1	0
1	1	0	1
1	1	1	1

A	B	C	Y
0	0	0	1
0	0	1	1
0	1	0	0
0	1	1	1
1	0	0	0
1	0	1	1
1	1	0	0
1	1	1	1

Figure A2.26

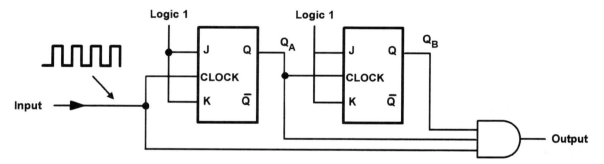

Figure A2.27

42. A sine wave oscillator is based on a Wien bridge with $R = 5$ kΩ and $C = 15$ nF. Determine the frequency of the output signal. [Page 162]

43. The frequency response characteristic of an operational amplifier is shown in Fig. A2.20. If the device is configured for a closed-loop gain of 200, determine the resulting bandwidth. [Page 150]

44. Redraw Fig. A2.21 using American (MIL/ANSI) symbols. [Page 175]

45. Draw the truth table for the logic gate arrangement shown in Fig. A2.22. [Page 175]

46. Redraw Fig. A2.23 using BS symbols. [Page 175]

47. What single logic gate can be used to replace the logic circuit shown in Fig. A2.24? [Page 176]

48. What single logic gate can be used to replace the logic circuit shown in Fig. A2.25? [Page 176]

49. Devise arrangements of logic gates that will produce the truth tables shown in Fig. A2.26. Use the minimum number of logic gates in each case. [Page 176]

50. A 1 kHz square wave clock waveform is applied to the circuit shown in Fig. A2.27. Sketch the output waveform against a labelled time axis. [Page 177]

Appendix 3

Answers to problems

Chapter 1
1.1 coloumbs, joules, hertz
1.2 3.6 MJ
1.3 0.52 radian
1.4 11.46°
1.5 39.57 kΩ
1.6 0.68 H
1.7 2.45 nF
1.8 0.19 mA
1.9 4.75×10^{-4} V
1.10 16.5×10^6 Ω
1.11 4.8×10^6, 7.2×10^3, 4×10^3, 0.5×10^{-3}
1.12 silver
1.13 33.3 mA
1.14 6.72 V
1.15 3.3 kΩ
1.16 15 Ω
1.17 0.436 Ω
1.18 0.029 W
1.19 0.675 W
1.20 57.7 mA
1.21 0.625×106 V/m
1.22 12 A
1.23 6 μWb

Chapter 2
2.1 60 Ω, 3.75 W, wirewound
2.2 270 kΩ 5%, 10 Ω 10%, 6.8 MΩ 5%, 0.39 Ω 5%, 2.2 kΩ 2%
2.3 44.65 Ω to 49.35 Ω
2.4 27 Ω and 33 Ω in series,
    27 Ω and 33 Ω in parallel,
    56 Ω and 68 Ω in series,
    27 Ω 33 Ω and 56 Ω in parallel,
    27 Ω 33 Ω and 68 Ω in series
2.5 66.67 Ω
2.6 10 Ω
2.7 102 Ω, 78.5 Ω
2.8 407.2 Ω
2.9 98.7 kΩ
2.10 3.21×10^{-4}
2.11 3.3 μF and 4.7 μF in parallel,
    1 μF and 10 μF in parallel,
    1 μF 3.3 μF 4.7 μF and 10 μF in parallel,
    1 μF and 10 μF in series,
    3.3 μF and 4.7 μF in series
2.12 60 pF, 360 pF
2.13 50 pF
2.14 20.79 mC
2.15 1.98 nF
2.16 69.4 μF
2.17 0.313 V
2.18 0.136 H
2.19 0.48 J
2.20 10 mH 22 mH and 60 mH in parallel,
    10 mH and 22 mH in parallel,
    10 mH and 22 mH in series,
    10 mH and 60 mH in series,
    10 mH 60 mH and 100 mH in series

Chapter 3
3.1 275 mA
3.2 200 Ω
3.3 1.5 A away from the junction, 215 mA away from the junction
3.4 1.856 V, 6.6 V
3.5 0.1884 A, 0.1608 A, 0.0276 A, 5.09 V, 0.91 V, 2.41 V
3.6 1.8 V, 10.2 V
3.7 0.5 A, 1.5 A
3.8 1 V, 2 V, 3 V, 4 V, 5 V
3.9 1.5 kΩ, 60 mA
3.10 40.6 ms
3.11 3.54 s
3.12 112.1 μF
3.13 0.128 s
3.14 2.625 V, 5 Ω
3.15 21 V, 7 Ω, 3 A, 7 Ω
3.16 50 mA, 10 V
3.17 50 V, 10 V

Chapter 4
4.1 4 ms, 35.35 V
4.2 59.88 Hz, 339.4 V
4.3 50 Hz 30 V pk-pk, 15 Hz 10 V pk-pk, 150 kHz 0.1 V pk-pk
4.4 19 V, −11.8 V
4.5 10.6 V

4.6   36.19 kΩ, 144.76 Ω
4.7   3.54 mA
4.8   10.362 Ω, 1.45 kΩ
4.9   4.71 V
4.10   592.5 Ω, 0.186 A
4.11   0.55, 0.487 A
4.12   157 nF
4.13   1.77 MHz to 7.58 MHz
4.14   7.5 mA, 2.71 V
4.15   281 kHz, 41.5, 6.77 kHz
4.16   18 V
4.17   245 V

Chapter 5
5.1   silicon (forward threshold appx. 0.6 V)
5.2   41 Ω, 150 Ω
5.4   9.1 V zener diode
5.5   250 Ω
5.6   germanium low-power high-frequency,
        silicon low-power low-frequency,
        silicon high-power low-frequency,
        silicon low-power high frequency
5.7   2.625 A, 20
5.8   5 mA, 19.6
5.9   16.7
5.10   BC108
5.11   8 μA, 1.1 mA
5.12   47 mA, 94, 75
5.13   12.5 mA, 12 V, 60 μA
5.14   16 mA

Chapter 6
6.1   80 mV
6.2   5 mV
6.3   200 Ω
6.4   12.74 V, 9.1 V, 8.4 V

6.5   36.4 mA, 0.33 W
6.6   0 V, 12.04 V, 0 V
6.7   0.5 Ω, 8.3 V
6.8   1%, 15.15 V

Chapter 7
7.1   40, 160, 6400, 100 Ω
7.2   2 V
7.3   56, 560 kHz, 15 Hz
7.4   18.5
7.5   0.0144
7.6   2.25 V
7.7   13 μA, 3.39 V, 2.7 V, 4.51 V
7.8   5 V, 7 mA, 8.5 V
7.9   12.2 V, 6.1 mA, 5.5 V

Chapter 8
8.1   10 V
8.2   40 dB, 600 kHz
8.3   200 kΩ
8.4   +1 V, −1 V, 0 V, −2 V, +2 V, 0 V
8.5   4 kHz, 100 Hz
8.6   11, 1
8.7   10, 3.38 kHz, 338 kHz

Chapter 9
9.1   4.44, 40
9.2   6.49 kΩ
9.3   18 kΩ
9.4   5.63 V pk-pk
9.5   14.3 kΩ, 42.9 kΩ

Chapter 10
10.9   Low-power Schottky (LS) TTL, 27th Month
        of 1989
10.10   0.6 V

Appendix 4

Semiconductor pin connections

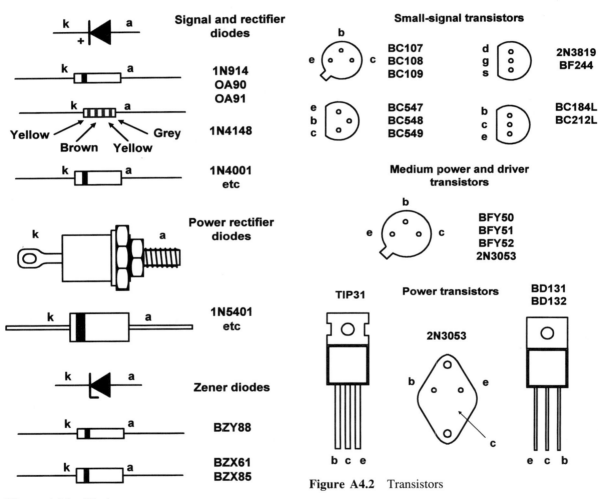

Signal and rectifier diodes

1N914
OA90
OA91

Yellow — Brown — Yellow — Grey

1N4148

1N4001
etc

Power rectifier diodes

1N5401
etc

Zener diodes

BZY88

BZX61
BZX85

Figure A4.1 Diodes

Small-signal transistors

BC107
BC108
BC109

2N3819
BF244

BC547
BC548
BC549

BC184L
BC212L

Medium power and driver transistors

BFY50
BFY51
BFY52
2N3053

Power transistors

TIP31

2N3053

BD131
BD132

b c e

e c b

Figure A4.2 Transistors

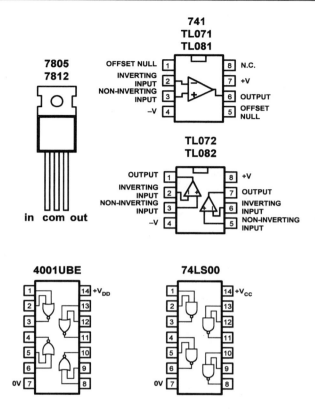

Figure A4.3   Integrated circuits

Appendix 5

Decibels

Decibels (dB) are a convenient means of expressing gain (amplification) and loss (attenuation) in electronic circuits. In this respect, they are used as a **relative** measure (i.e. comparing one voltage with another, one current with another, or one power with another). In conjunction with other units, decibels are sometimes also used as an **absolute** measure. Hence dBV are decibels relative to 1 V, dBm are decibels relative to 1 mW, etc.

The decibel is one-tenth of a bel which, in turn, is defined as the logarithm, to the base 10, of the ratio of output power (P_{out}) to input power (P_{in}).

Gain and loss may be expressed in terms of power, voltage and current such that:

$$A_P = \frac{P_{out}}{P_{in}} \quad A_V = \frac{V_{out}}{V_{in}} \quad \text{and } A_I = \frac{I_{out}}{I_{in}}$$

where A_P, A_V or A_I is the power, voltage or current gain (or loss) expressed as a ratio, P_{in} and P_{out} are the input and output powers, V_{in} and V_{out} are the input and output voltages, and I_{in} and I_{out} are the input and output currents. Note, however, that the powers, voltages or currents should be expressed in the same units/multiples (e.g. P_{in} and P_{out} should both be expressed in W, mW, μW or nW).

It is often more convenient to express gain in decibels (rather than as a simple ratio) using the following relationships:

$$A_P = 10 \log_{10}\frac{(P_{out})}{(P_{in})} \quad A_V = 20 \log_{10}\frac{(V_{out})}{(V_{in})}$$

$$\text{and } A_I = 20 \log_{10}\frac{(I_{out})}{(I_{in})}$$

Note that a positive result will be obtained whenever P_{out}, V_{out}, or I_{out} is greater than P_{in}, V_{in}, or I_{in}, respectively. A negative result will be obtained whenever P_{out}, V_{out}, or I_{out} is less than P_{in}, V_{in} or I_{in}. A negative result denotes attenuation rather than amplification. A negative gain is thus equivalent to an attenuation (or loss). If desired, the formulae may be adapted to produce a positive result for attenuation simply by inverting the ratios, as shown below:

$$A_P = 10 \log_{10}\frac{(P_{in})}{(P_{out})} \quad A_V = 20 \log_{10}\frac{(V_{in})}{(V_{out})}$$

$$\text{and } A_I = 20 \log_{10}\frac{(I_{in})}{(I_{out})}$$

where A_P, A_V or A_I is the power, voltage or current gain (or loss) expressed in decibels, P_{in} and P_{out} are the input and output powers, V_{in} and V_{out} are the input and output voltages, and I_{in} and I_{out} are the input and output currents. Note, again, that the powers, voltages or currents should be expressed in the same units/multiples (e.g. P_{in} and P_{out} should both be expressed in W, mW, μW or nW).

It is worth noting that, for identical decibel values, the values of voltage and current gain can be found by taking the square root of the corresponding value of power gain. As an example, a voltage gain of 20 dB results from a voltage ratio of 10 while a power gain of 20 dB corresponds to a power ratio of 100.

Finally, it is essential to note that the formulae for voltage and current gain are only meaningful when the input and output impedances (or resistances) are identical. Voltage and current gains expressed in decibels are thus only valid for matched (constant impedance) systems.

The following table gives some useful decibel values:

Decibels (dB)	Power gain (ratio)	Voltage gain (ratio)	Current gain (ratio)
0	1	1	1
1	1.26	1.12	1.12
2	1.58	1.26	1.26
3	2	1.41	1.41
4	2.51	1.58	1.58
5	3.16	1.78	1.78
6	3.98	2	2
7	5.01	2.24	2.24
8	6.31	2.51	2.51
9	7.94	2.82	2.82
10	10	3.16	3.16
13	19.95	3.98	3.98
16	39.81	6.31	6.31

Decibels (dB)	Power gain (ratio)	Voltage gain (ratio)	Current gain (ratio)
20	100	10	10
30	1000	31.62	31.62
40	10 000	100	100
50	100 000	316.23	316.23
60	1 000 000	1000	1000
70	10 000 000	3162.3	3162.3

Example A5.1

An amplifier with matched input and output resistances provides an output voltage of 1 V for an input of 25 mV. Express the voltage gain of the amplifier in decibels.

Solution

The voltage gain can be determined from the formula:

$A_V = 20 \log_{10}(V_{out}/V_{in})$

where $V_{in} = 25$ mV and $V_{out} = 1$ V.
   Thus:

$A_V = 20 \log_{10}(1 \text{ V}/25 \text{ mV}) = 20 \log_{10}(40)$
$\quad = 20 \times 1.6 = 32$ dB

Example A5.2

A matched 600 Ω attenuator produces an output of 1 mV when an input of 20 mV is applied. Determine the attenuation in decibels.

Solution

The attenuation can be determined by applying the formula:

$A_V = 20 \log_{10}(V_{in}/V_{out})$

where $V_{in} = 20$ mV and $V_{out} = 1$ mV.
   Thus:

$A_V = 20 \log_{10}(20 \text{ mV}/1 \text{ mV}) = 20 \log_{10}(20)$
$\quad = 20 \times 1.3 = 26$ dB

Example A5.3

An amplifier provides a power gain of 33 dB. What output power will be produced if an input of 2 mW is applied?

Solution

Here we must re-arrange the formula to make P_{out} the subject, as follows:

$A_P = 10 \log_{10}(P_{out}/P_{in})$

thus

$A_P/10 = \log_{10}(P_{out}/P_{in})$

or

$\text{antilog}_{10}(A_P/10) = P_{out}/P_{in}$

Hence

$P_{out} = P_{in} \times \text{antilog}_{10}(A_P/10)$

Now $P_{in} = 2$ mW $= 20 \times 10^{-3}$ W and $A_P = 33$ dB, thus

$P_{out} = 2 \times 10^{-3} \times \text{antilog}_{10}(33/10)$
$\quad = 2 \times 10^{-3} \times \text{antilog}_{10}(3.3)$
$\quad = 2 \times 10^{-3} \times 1.995 \times 10^{-3} = 3.99$ W

Index

3dB points, 125

Absolute permeability, 13
AC coupled amplifier, 121
Acceptor circuit, 76
Air-cored inductor, 39, 45
Alternating current, 68
Alternating voltage, 68
Aluminium, 7
Ambient temperature, 19, 34
Amp, 1, 3
Amplifier, 121, 141
Amplifier circuits, 136, 137, 139, 140
AND gate, 175, 186
AND logic, 173
Angle, 2
Anode, 84
ANSI logic symbol, 175
Apparent power, 74
Assignments, 190, 192
Astable multivibrator, 163, 164, 168, 191
Astable oscillator, 165, 166, 169
Atom, 83
Audio frequency amplifier, 121, 126
Avalanche diode, 87
Average value, 70

B-H curve, 14, 15
Back e.m.f., 39, 41
Balance, 52
Bandwidth, 76, 77, 125, 126, 127, 148, 149, 150
Base, 95
Bi-phase rectifier, 113, 114, 115
Bias, 123, 125, 135
Bias stabilization, 140
Bias voltage, 96
Binary counter, 178, 179
Binary logic, 173
Bipolar transistor, 93
Bistable, 177
Bistable latch, 178
Bistable multivibrator, 163
Bobbin, 78
Bonding, 83
Bridge rectifier, 114, 115, 116, 119
BS logic symbol, 175
BS 1582, 22
Buffer, 174, 186

C-R circuit, 55, 56, 57, 59
C-R ladder network, 161
Cadmium selenide, 29

Cadmium sulphide, 29
Capacitance, 1, 32, 33, 34
Capacitive reactance, 71
Capacitor, 30, 31
Capacitor markings, 34
Capacitor specifications, 34
Carbon, 28
Carbon film resistor, 19
Carbon rod resistor, 19
Cast iron, 14
Cast steel, 14
Cathode, 84
centi, 4
Centre-zero meter, 65
Ceramic capacitor, 34
Ceramic wirewound resistor, 19
Charge, 1, 9, 31, 33
Charge carrier, 83
Charging circuit, 55
Chassis, 16
Circuit symbols, 16, 45, 65, 80, 106, 119, 141, 146,
    186
Class A, 122, 135, 136
Class AB, 123
Class B, 124
Class C, 125
Class of operation, 122, 125
Clear, 177
Clipping, 123
Clock, 177
Clock frequency, 184
Closed-loop gain, 128
Closed-loop voltage gain, 147
CMOS, 179, 183, 184
CMOS device coding, 183
CMRR, 149, 155
Collector, 95
Colour code, 21, 22, 35
Combinational logic, 176
Common base, 131
Common collector, 129, 131
Common drain, 129, 130, 131
Common emitter, 97, 129, 130, 131, 132, 133, 134,
    138, 140
Common gate, 130, 131
Common mode rejection, 149, 155
Common source, 129, 131
Complementary metal oxide semiconductor, 179
Component symbols, 16, 45, 65, 80, 106, 119, 141,
    146, 186
Conductivity, 7
Conductor, 5

Constant current source, 65
Copper, 7, 28
Coulomb, 1, 2, 3
Counter, 179
Covalent bonding, 83
Crystal, 170
Crystal oscillator, 166
Current, 1, 6, 13
Current Law, 48
Current divider, 51
Current gain, 97, 99, 121, 131, 132, 205
Current limiting resistor, 93
Cut-off, 123, 135
Cut-off frequency, 125

D-type bistable, 177, 178, 187
Date code, 180
DC coupled amplifier, 121
DC level, 68
Decibels, 205
Depletion layer, 84, 85
Derived units, 1
Diac, 91, 92, 106
Dielectric, 30, 33, 35
Differential amplifier, 153
Differentiating circuit, 59, 60, 63
Digital integrated circuit, 104
Digital logic, 172
DIL package, 105
Diode, 84, 85, 87, 106
Diode characteristic, 86
Diode coding, 89, 90
Diode packages, 90
Diode test circuit, 87
Discharge circuit, 57
Doping, 84
Dual-in-line package, 105
Dynamic power dissipation, 184

E6 series, 18, 20
E12 series, 18, 20
E24 series, 18, 20
Earth, 16
Electric charge, 9
Electric field, 9, 10
Electric field strength, 10
Electrolytic capacitor, 34, 45
Electromagnetism, 10
Electromotive force, 5, 6
Electron, 5, 83
Electron flow, 31, 32, 36
Electrostatics, 9
E.m.f., 5, 6, 13
Emitter, 95
Emitter follower, 141
Energy, 1, 8, 33
Energy storage, 33, 39
Equivalent circuit, 124, 130, 131, 134
Exclusive-OR gate, 176

Exponential decay curve, 58, 59, 62
Exponential growth curve, 56, 59, 61
Exponents, 4

Fan-out, 184
Farad, 3, 33
Feedback, 128
Ferrite cored inductor, 42, 45
Ferrite cored transformer, 80
Ferrite pot cored inductor, 42
FET, 101
FET characteristics, 104
FET parameters, 102, 103
Field effect transistor, 101
Fixed capacitor, 45
Fixed inductor, 45
Fixed resistor, 16
Flux, 11, 13
Flux collapse, 40, 41
Flux density, 11, 14, 15
Flywheel action, 124
Force, 1
Force between charges, 9
Force between conductors, 10
Formulae, 16, 44, 64, 80, 106, 142, 155, 170, 187
Forward bias, 84, 85
Forward current, 86
Forward threshold voltage, 84, 85
Forward transfer conductance, 102
Forward voltage, 86
Four terminal network, 141
Free electron, 83
Free space, 9, 10, 11, 33
Free running multivibrator, 163
Frequency, 1, 69
Frequency range, 42
Frequency response, 125, 126, 150, 154
Frequency response tailoring, 152
Fringing, 10, 12, 14
Full-power bandwidth, 148
Full-wave rectifier, 113
Fundamental, 127
Fundamental mode, 167
Fundamental units, 1

Gain, 149, 150
Gate, 91
Germanium, 86, 95
Giga, 4
Ground, 16

h-parameters, 99, 130, 132, 134
Half-wave rectifier, 110, 111, 112, 191
Harmonic, 127
Heat, 8
Heat-sensing unit, 191
Helium, 5
Henry, 1, 3
Hertz, 1, 3

Hole, 84
Hybrid parameters, 130, 131

IC voltage regulator, 118, 119
Ideal amplifier characteristics, 150
IGFET, 101
Illuminance, 1
Impedance, 73
Impedance triangle, 74
Impurity, 83
Induced voltage, 39
Inductance, 1, 39, 41
Inductive reactance, 72
Inductor markings, 42
Inductor specifications, 42
Inductors, 38
Input bias current, 157
Input characteristic, 96, 132, 133
Input impedance, 124
Input offset voltage, 148
Input resistance, 124, 125, 130, 131, 132, 147, 153, 155
Insulated gate field effect transistor, 101
Insulator, 5
Integrated circuit packages, 105
Integrated circuits, 104
Integrating circuit, 59, 60, 63
Internal gain, 128, 147, 160
Internal resistance, 117
Inverter, 175, 186
Inverting amplifier, 151, 152, 155
Inverting input, 146
Iron, 28
Iron-cored inductor, 42, 45
Iron-cored transformer, 80

J-K bistable, 177, 178, 179, 180, 187
JFET, 101
Joule, 1, 2, 3, 8
Junction, 16
Junction diode, 84
Junction gate FET, 101, 106

kilo, 4
Kirchhoff's Laws, 48

L-C circuit, 75
L-C-R circuit, 75
L-R circuit, 61, 63
Ladder network, 166
Ladder network oscillator, 161, 170
Lamination, 78
Large-signal amplifier, 121
LDR, 29, 45
LED, 93
Lattice, 83
Lead, 7
Leakage current, 84
Leakage flux, 12
Length, 1

Light, 8
Light-dependent resistor, 29, 45
Light-emitting diode, 93, 106
Light-operated switch, 190
Linear amplifier, 122, 135
Linear integrated circuits, 104
Linear law potentiometer, 30
Load line, 137, 138
Loads, 18
Logarithmic law potentiometer, 30
Logic 0, 184
Logic 1, 184
Logic circuit characteristics, 183
Logic circuits, 172
Logic family, 179, 184
Logic family characteristics, 184
Logic gate packages, 185
Logic gate, 174
Logic level, 183, 184
Loop gain, 128, 161, 170
Low-noise amplifier, 121
Lower cut-off frequency, 157
Lumen, 1
Luminous flux, 1
Luminous intensity, 1
Lux, 1

Magnetic circuit, 12, 13, 15
Magnetic field, 10, 11, 12
Magnetic field strength, 11
Magnetic flux, 1, 39
Magnetic flux density, 11
Magnetic pole, 10
Magnetic saturation, 15
Magnetizing force, 14
Magnetomotive force, 13
Main terminal, 91, 92
Mass, 1
Matter, 1
Maximum reverse repetitive voltage, 86
Mega, 4
Metal film resistor, 19
Metal oxide resistor, 19
Metallized film capacitor, 34
Metals, 7
Meter, 65
Mica capacitor, 34
micro, 4
Mid-band, 124
MIL symbol, 175
Mild steel, 7, 14
milli, 4
M.m.f., 13
Monostable multivibrator, 163
Multi-cell battery, 16
Multi-plate capacitor, 34
Multiples, 3
Multivibrator, 163
Mutual characteristic, 102

N-type material, 84
NAND gate, 175, 182, 183, 186
nano, 4
Negative charge carriers, 5
Negative feedback, 127, 128, 136
Negative temperature coefficient, 25, 28
Newton, 1
Noise margin, 183, 184, 185
Noise performance, 19
Non-inverting amplifier, 153, 155
Non-inverting input, 146
Non-linear amplifier, 122
Norton's theorem, 54
NOR gate, 175, 186
NOT gate, 186
NPN transistor, 95, 96, 106
N.t.c., 25, 28
N.t.c. thermistor, 45
Nucleus, 5

Offset-null, 148
Ohm, 1, 3
Ohm's Law, 6
Open-loop frequency response, 150
Open-loop voltage gain, 147, 155, 157
Operational amplifier, 157
Operational amplifier characteristics, 149, 150, 153, 157
Operational amplifier circuits, 152
Operational amplifier packages, 156
Operational amplifier types, 154
Operational amplifiers, 146
OR gate, 175, 186
OR logic, 173
Oscillation, 161
Oscillator circuits, 167
Oscillators, 160
Output characteristic, 97, 103
Output impedance, 124
Output resistance, 117, 125, 131, 148, 155
Output voltage swing, 149
Over-driven amplifier, 122
Overtone operation, 167

P-type material, 84
Parallel connected capacitors, 35, 37
Parallel connected inductors, 42, 43
Parallel connected resistors, 23, 27
Parallel plate capacitor, 30
Parallel resonant circuit, 75
Passive components, 18
P.d., 5
Peak inverse voltage, 86
Peak value, 70
Peak-peak value, 70
Pentavalent impurity, 84
Period, 70
Periodic time, 69, 70, 170
Permeability, 12, 13, 14, 15, 41
Permeability of free space, 10, 11

Permittivity of free space, 9
Phase angle, 74
Phase shift, 127, 131, 153
Phasor diagram, 72, 73
Photodiode, 106
pico, 4
Piezoelectric effect, 167
Pin connections, 203
PIV, 86
Platinum, 28
PNP transistor, 95, 96, 106
Polyester capacitor, 34
Polystyrene capacitor, 35
Positive charge carriers, 5
Positive feedback, 160
Positive temperature coefficient, 25, 28
Pot cored inductor, 42
Potential, 1
Potential difference, 5, 6
Potential divider, 50
Potentiometer, 30, 45
Power, 1, 8
Power factor, 74
Power gain, 131, 205
Power rating, 19, 20
Power supply, 109, 191
Power supply circuits, 118
Preferred values, 18
Preset, 177
Preset capacitor, 45
Preset inductor, 45
Preset potentiometer, 45
Primary winding, 78, 79
Problems, 17, 46, 64, 81, 106, 119, 142, 157, 170, 187, 193
Propagation delay, 184
Proton, 5
P.t.c., 25, 28, 29
P.t.c. thermistor, 45
Push-button switch, 65
Push-pull, 124

Q-factor, 42, 76, 77
QIL package, 105
Quad-in-line package, 105
Quality factor, 42, 76
Quantity of electricity, 33
Quartz crystal, 170, 167

R-S bistable, 177, 187
Radian, 2
Radio-frequency amplifier, 126
Ratio arms, 52
Reactance, 71, 72, 73, 74
Real amplifier characteristics, 150
Rectifier, 109, 110, 113, 114
Rectifier diode, 87
Regulation, 117
Rejector circuit, 76

Relative conductivity, 7
Relative permeability, 13
Relative permittivity, 33
Reluctance, 12, 13
Reservoir, 109, 111, 114
Reset, 177
Resistance, 1, 5, 6, 7, 13
Resistance versus temperature, 25, 27
Resistivity, 7
Resistor, 19
Resistor markings, 21
Resistors, 18
Resonance, 76
Reverse bias, 84, 85
Reverse breakdown, 87
Reverse breakdown voltage, 84
Reverse voltage, 84, 86
Revision problems, 193
Righ-hand screw rule, 11
Ripple, 111, 112
R.m.s. value, 70
Rotary potentiometer, 30

Saturation, 135
Scientific notation, 4
Second, 3
Secondary winding, 78, 79
Semiconductor, 83
Semiconductor diode, 84
Semiconductors, 203
Semi-logarithmic law potentiometer, 30
Series connected capacitors, 35, 37
Series connected inductors, 42, 43
Series connected resistors, 23, 26
Series regulator, 109
Series resonant circuit, 75
Series-pass transistor, 118, 119
Set, 177
Shell, 5
Shift register, 180, 181
Shunt resistor, 51
SI units, 1
Siemen, 3
Signal diode, 87
Signals, 68, 69
SIL package, 105
Silicon, 83, 86, 95
Silicon controlled rectifier, 89
Silicon steel, 14
Silver, 7, 28
Sine wave, 2, 68
Single-cell battery, 16
Single-in-line package, 105
Slew rate, 148, 155, 157
Small-signal amplifier, 121
Smoothing, 111, 112, 113
Smoothing filter, 109
Solenoid, 12
Sound, 8

Source follower, 129, 130
SPDT switch, 65, 119
Specific resistance, 7
Specifications, 34
Speed-power product, 184
SPST switch, 65
Stability, 19, 34
Static power dissipation, 184
Steel, 14
Step-down transformer, 109
Student assignments, 190
Sub-multiples, 3
Switch, 65
Switch logic, 172
Symbols, 16, 45, 65, 80, 106, 119, 141, 146, 186
Symmetry, 152

Temperature, 1
Temperature coefficient, 7, 19, 25, 27, 34
Tera, 4
Terminal, 65
Tesla, 1, 3, 11
Test circuit, 87, 89, 100, 104
Thermistor, 28
Thevenin's theorem, 52
Threshold voltage, 84, 166
Thyristor, 89, 91, 106
Time, 1
Timing diagram, 178, 180, 181
Tolerance, 18, 19, 34, 42
Transfer characteristic, 98, 99, 134
Transformer, 77, 79
Transistor, 95
Transistor amplifier, 129
Transistor amplifier configurations, 129
Transistor characteristics, 100
Transistor circuit configurations, 131
Transistor coding, 94
Transistor packages, 104, 105
Transistor parameters, 99
Transistor transistor logic, 179
Transistor types, 94
Transition frequency, 99
Triac, 90, 92, 106
Triangle wave generator, 169
Triggering, 90
Trivalent impurity, 84
True power, 74
Truth table, 173, 174, 175
TTL, 179, 182, 184
TTL device coding, 183
Turns ratio, 78, 79

Units, 1
Upper cut-off frequency, 155

Valence shell, 83
Variable capacitance diode, 89, 91, 106
Variable capacitor, 38, 45

Variable inductor, 43, 45
Variable resistor, 29, 30, 45
Varicap, 89
VDR, 29, 45
Vitreous wirewound resistor, 19
Volt, 1, 2, 3, 8
Voltage, 5
Voltage Law, 48
Voltage dependent resistor, 29, 45
Voltage gain, 121, 131, 135, 153, 205
Voltage rating, 34
Voltage ratio, 78
Voltage reference, 109
Voltage regulator, 109, 116, 117

Watt, 2, 3, 8
Waveforms, 60, 63, 64, 68, 69
Waveshaping, 59, 60, 63
Weber, 1, 3
Wheatstone bridge, 52, 53
Wideband amplifier, 121, 126
Wien bridge oscillator, 162, 163, 170, 168
Wirewound resistor, 19

XNOR gate, 186
XOR gate, 186

Zener diode, 87, 88, 89, 106, 117
Zener diode regulator, 118